The new equine economy in the 21st century

EAAP – European Federation of Animal Science

The European Association for Animal Production wishes to express its appreciation to the *Ministero per le Politiche Agricole e Forestali* and the *Associazione Italiana Allevatori* for their valuable support of its activities

The new equine economy in the 21st century

EAAP publication No. 136

edited by:

Céline Vial

Rhys Evans

EAN: 9789086862795
e-EAN: 9789086868247
ISBN: 978-90-8686-279-5
e-ISBN: 978-90-8686-824-7
DOI: 10.3920/978-90-8686-824-7

ISSN 0071-2477

Photo cover: Ann Kern-Godal,
Oslo University Hospital

First published, 2015

©Wageningen Academic Publishers
The Netherlands, 2015

Wageningen Academic Publishers
P.O. Box 220
6700 AE Wageningen
The Netherlands
www.WageningenAcademic.com
copyright@WageningenAcademic.com

Table of contents

Part 1.
Introduction

Introduction to the new equine economy in the 21st century

R. Evans

Norwegian University College for Agriculture and Rural Development (Hogskulen for landbruk og bygdeutvikling – HLB), 4353 Klepp Stasjon, Norway; rhys@hlb.no

The development and transformation of the European horse industry

The world of horses and riding has undergone significant changes since the middle of the 20th century. But then, so has human society. And the changes in the equine sector reflect some of the wider changes in society. When we talk about economics, after all, we are talking about a *social* science – one in which human diversity, change and indeed, perversity must be taken into account in order to understand that most socially constructed of phenomena – value. As European and other societies have changed, so too, the value of horses has changed in response. A classic example comes from the period immediately prior to the one in which we are interested – the value of horses to farmers. Prior to the growth of the internal combustion engine, horses were the most valuable asset most farmers could aspire to possessing. And many cultural practices, from breeders creating the most efficient working horses, to the creation of elaborate decorative horse carriages, were spawned from the high value represented by horses in a pre-internal combustion engine regime. This was all part of an equine economy which generated high levels of economic activity across society.

This, however, all changed after WWII (World War II). The impact of the 'first mechanized war' helped to increase the spread of internal combustion engines, powering tractors, trucks, feed mills, etc. This was catastrophic for horses themselves (estimates generally agree that horse numbers declined approximately 90% in Europe by the 1950s) and for the equine economy (EHN, 2014). Prior to this, 90% of the horse population was generally employed in agriculture, industry or transportation, and the 10% that remained were kept for reasons of affection, tradition, sport and in some cases, leisure. A key new concept (where horses are concerned) is the idea that there must be an *'economic imperative'* for breeding them (Evans, 2015). In other words, when horses (or for that matter, other things) are part of an active economic sector (such as farm production), there is an economic incentive to increase the quantity and improve their quality. When that incentive is removed, the result is a lack of breeding and the type of demographic declines we have witnessed in the mid-20th century. Now we are looking at a resurgent population, based upon new activities, and new economic imperatives. Given this, we can say that with the transformation of the means of production came a transformation of the purposes to which horses were put. And so the basis for the new equine economy in the 21st century was laid.

Today in Europe, it is estimated that there could be at least 6 million horses in the 27 country members, grazing 6 million hectares of permanent grassland. Four hundred thousand full time jobs equivalent are estimated to be provided by the sector and the numbers of horses and riders are growing in the approximate range of 7% a year (EHN, 2014). Despite this, remarkably little is known about the horse industry and its development, at both national and international levels. Few studies have been conducted and in most countries, statistical data on the horse sector is lacking or incomplete.

Members of the Horse Commission have documented the regular annual increase in ridership and horse numbers in Europe in the last decade and a half, and that increase is accompanied by some important new factors. These include the growth of ridership in peri-urban areas, the increased gendering of the sector (well over 70% female ridership across many European nations), and the rise of the popularity of horse riding in the face of increasing urbanization. Other issues addressed by our members include the use of horses in sustainable management of landscapes, issues of horse welfare, and, of course, innovations in veterinary science and ethology. Across the Horse Commission we deal with contemporary issues, using our scientific expertise to address new issues which are constantly arising from these new formations of the horse industry. This book is the result of those studies.

Human-horse relations at the centre of the new equine economy

This book sits within a relatively new field of studies which can be called 'Human-horse relations'. Historically, horses have been studied within different disciplines of equine science, and many studies (with a few notable exceptions) have focused upon the species in isolation from the social context within which they are situated. There has been much excellent science generated which focuses upon a limited number of use contexts, such as sport, or indeed, the consequences of common stable practices. But many of these studies take these contexts as given categories, with the focus of exploration on the biological processes of the animal. Human-horse relations research begins by acknowledging that horses predominantly exist within a human-framed context and from this, explores that context, its consequences (for horses and for humans), and attempts to understand how both human practices affect horses, and how horse ethology and biology affects what humans can, and can't do with them. Given that the overwhelming proportion of horses are not now used in sport, or in traditional 'horse work', our lack of scientific understanding of the nuances of human-horse relations has profound consequences for policy questions ('Is the horse an agricultural animal?'), for our understanding of the impact of equine activities on the built and natural environment, and for horse welfare itself. The field of human-horse relations includes considerations of human society (economics, demographics, gender, etc.), of equine ethology and biology (i.e. the debate about 'free stalling', or 'natural' verses 'traditional' horsemanship, etc.), how the social (and personal) construction of the meaning of horses in human life affects how we treat them, and, importantly, whether their numbers flourish or decline.

The field of study has also been called 'socio-economic' studies of horses, which is a good beginning. Here, it is important to see economics as a social science, for this combination of the study of society and of economics forms a strong foundation for building our understanding of human-horse relations. However, the term 'economics' traditionally encourages us to limit our focus quite narrowly on issues of economic valuation. If we are to understand the factors which drive the growth of this economic sector, however, we must also acknowledge the fundamental passion that folk have for horses, and this is driven by other factors, such as cultural and historical practice, and meanings. To understand the passion for horses that drives the recent economic growth of the sector, we must also understand a bit of what happens *inside* the riders, and *inside* the horses too. We can do this by looking more deeply into the nature of the human-horse bond, as this is an extremely important motivation driving the growth of the sector. And so, the study of human-horse relations rigorously examines multiple factors, from multiple disciplines in an attempt to understand how we got here, and where the sector will go in the future.

The new equine economy of the 21st century

Given the above, it is obvious that in order to understand the new equine economy in the 21st century, we need first to understand some of the changes taking place across the wider human society within which horses are located. Horses are no longer a part of most European 'production' processes, whether on the farm, on the roads, or in industry. How and why, then, is the sector growing as it appears to be, across much of Europe, and what might the consequences of this new growth be for both horses and people, as well as local areal and rural development?

To understand the new equine economy in the 21st century, we need to understand the wider economy within which horses and riders are located. In particular, as the 21st century with its 'Knowledge Economy' becomes established, we see what appears to be an ever-increasing trend towards urbanization across Europe – more and more people live in cities. These urban lives can be characterized by the replacement of social bonds with technological ones, by the separation of domestic residence and workplace, and by growing levels of stress and the challenges to wellbeing as a result of increasingly sedentary lives, as well as by a relative increase in prosperity among the individuals involved. In a globalized economy where it seems that places are becoming more and more alike and where surveillance increasingly penetrates both public and private lives, there appears to be a growing need to balance all this with something which allows individuals to take contact with the world in which human beings evolved – a world of 'nature', of phenomena such as weather, and of the types of interactions (between human and world) for which our organism has evolved for hundreds of thousands of years.

Horses provide one way of rebalancing increasingly urban lives. To increasingly highly-stressed urban citizens, horses offer five key activities:
1. daily 'embodied' outdoor activity and encounters with 'nature';
2. regular physical exercise which can reduce stress, increase fitness and support greater wellbeing;
3. experience of 'authenticity' in a world of artifice;
4. ineffable contact with a large, non-human 'other' and the potential transcendence of individual isolation;
5. social stimulation with both humans and non-human others.

For many people, access to these phenomena does not come from their education or employment. Thus, they look to their leisure time to satisfy them. The leisure use of horses has always existed, but in the 21st century it has become increasingly predominant in terms of overall levels of activities within the sector. Riding and horse keeping is overwhelmingly a leisure activity now. And, within the wider society, leisure activities is now a significant (some would argue the *most significant*) sector of economic activity – and one which continues to grow. Thus, we must view our love of horses and the growth of riding as a component of one of the most significant economic sectors in modern times.

Just like other leisure activities, markets in the equine sector are influenced by what people believe, by what they are told is important, and by their own drives to meet their own needs. Leisure is, on one hand, an optional activity which we indulge in when not working. On the other hand, it is also necessary to reproduce our labour power – by keeping us fit and healthy, and providing necessary relief from the stresses of our urban lives (Figure 1). As such, it sits between being a necessity and a luxury. Or perhaps it is both. In any case, in order to understand the diverse uses to which humans put horses, we must understand both the way in which human society is changing and the fundamental nature of the human-horse relationship.

There are many factors to consider here. And the authors in this book consider many of them. They range from social and cultural factors, to economic and environmental ones. And, whilst horses are not a major factor in agriculture, they do remain an important component of rural development.

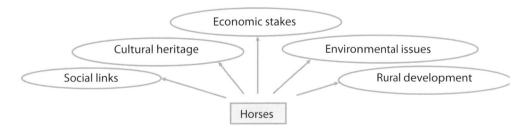

Figure 1. The complex set of factors influencing horse use in the 21st century.

If we can understand these underlying factors, we can begin to approach an understanding of the shape and potential of the new equine economy in the 21st century.

This book

We have called this book 'The new equine economy in the 21st century' because we wish to call formal attention to recent changes in society's use of horses and its valuing of them. The relationship between a society's values and its economy are inevitably entwined, and this is expressed very strongly in the European equine sector. This book represents a first attempt to come to grips with these changes and to speculate on what they mean for the sector.

Many European nations, for example, now do not classify horses as 'Agricultural animals'. In others, of course, horses are raised as food. But in many of the former, the numbers of horses and riders is growing annually, despite horses not being part of the farming production process any more. In most nations, the leisure sector (including recreation, fitness and wellbeing, tourism, etc.) has grown until it is one of the dominant sectors of economic activity. Horse keeping and riding now primarily occurs as part of this leisure sector, and yet attitudes, and policies, are still catching up to these changes.

There is increasing recognition that the status of horses in Europe is one of renaissance after serious decline and, like any new social phenomenon, this renaissance brings with it new formations, new questions and new opportunities. This book, therefore, is a collection of tightly focused observations on this change, by authors who have undertaken mostly qualitative social science research on the aspect of the new formations they write about. It is an outcome of a session held at the 2013 Annual Congress of the European Association of Animal Science in Nantes, France, by members of the Horse Commission. The session was also called 'The new equine economy in the 21st century.' It was created by the two editors of this book and the majority of the chapters here are written by participants in this session.

The book is divided into eight sub-sections, each focussing on a specific aspect of the new equine economy of the 21st century. The first focuses on specific economic research into aspects of the sector. Subsequent sections focus upon the opportunities and challenges it raises. One section explores how horse people participate in a social economy through new social media and new understandings of alternative health care. Another focuses on the current situation and new challenges this new formation has for governance of the sector. The final two sub-sections specifically focus on two new aspects – the use of horses as components of systems of sustainable grazing management, and the scientific raising of horses as food. The latter marks the first time such a comprehensive guide to all aspects of horse husbandry has been published and points the way to creating a system of food production which meets contemporary animal welfare and environmental sustainability considerations whilst delivering a healthy protein alternative to growing population and demands for such.

Conclusion

The relationship between people and horses is always evolving as society evolves. At its core, however, is a passion for horses built upon Xenophon's prescient observation thousands of years ago that 'Horses are already half human, and humans already half horse' (Xenophon, 1962). This book explores many aspects of that passion as a basis for creating an understanding of its economic consequences. And, based upon formal scientific rigour, the authors illuminate both where we have got to now in our relationship with horses, and where that relationship might go in the future.

These, then, are some of the varied aspects of the new equine economy in the 21st century. It turns out that the equine economy is as varied as the places within which it is situated, across a long continuum of practices. What is clear, however, is that just as the wider economy in Europe has moved from being dominated by industrial production to the consumption and leisure services, so too, the equine economy is making such a diversification. By understanding this new equine economy better, we can inform better policy choices, improve the way we educate young potential equine entrepreneurs, and support those already in the field to innovate and make their businesses truly sustainable. By doing this, we will be helping establish new economic imperatives for the breeding and production of horses, and in so doing we will better be able to support the flourishing of the new equine economy in the 21st century.

References

EHN, 2014. European horse network. Available at: www.europeanhorsenetwork.eu/the-horse-industry/economic-impact.
Evans, R., 2015. Riding native Nordic breeds: native breeds as cultural heritage development in Nordic countries. In: Evans, R. and S. Pickel Chevalier (eds.) Cheval, tourisme et sociétés. Mondes du Tourisme, Angers, France.
Xenophon, 1962. The art of horsemanship. J.A. Allen, London, United Kingdom, 187 pp.

Part 2.

Impact studies

This part of the book focuses on the economic impact of horse-related activities at fairly large scales in various locales. This is clearly an important issue. When one organisation wishes to promote the sector's importance in Europe, for example, they use claims that it generates '100 billion euros a year' in economic activity (EHN, 2014[1]). Of course, they – and anyone else – can only work with the figures that exist, and every national survey seen by this author claims that there is at least 25% under reporting due to reliance on equine and agricultural organisations for information. Horses which are not registered in these organisations do not show up on horse population figures. Any research which attempts to capture such large figures requires accurate research into smaller, specifically-located impacts and a credible system to sum them up. Clearly '100 billion euros' is a general estimate and there is a need for more accuracy for good policy decisions to be made at national and European levels. What is needed is a complete set of detail research at local and regional levels which can be scaled up to nation and international levels. Moreover, impact studies can focus on the whole horse industry of a region or on some activities or sectors, such as equestrian events, equestrian tourism, races, etc. They often concern economic effects but social and environmental aspects are also important to study and take into account to promote a sustainable development of the horse industry. These authors have begun that task.

The first three authors specifically look at the equine economies of France, of Norway/Sweden, and of Kentucky in the USA. Even so, the first two necessarily focus on specific events and locations in order to estimate or extrapolate the real economic impact of equestrian activities at these large scales. This highlights a key gap in the research – there are very few comprehensive national studies which address the important issue of just how much economic activity is generated by the growing equine sector at this level. It also points to another key issue, which is agreement about the methodology needed to capture this activity at such a large scale, including estimates of the multiplier effects of direct activity. It is clear that there is a strong need to know how many horses and how much economic activity is generated with them at a European and national scale in order to inform policy which supports and manages such activity. The third paper, which reports on the Kentucky Equine Survey offers one schema for doing this and could be used as a spring board for the further development of a methodology for capturing the overall economic impacts of equine activities.

---

[1] European Horse Network, 2014. Available at: http://tinyurl.com/ot5rwlv.

1. Economic impact of equestrian events, examples from France

C. Vial[1,2]*, E. Barget[3] and F. Clipet[2,3]

[1]French Institute for Horse and Riding, IFCE, 19000 Arnac-Pompadour, France; vialc@supagro.inra.fr
[2]National Institute of Agronomic Research, INRA, UMR 1110 MOISA, 34000 Montpellier, France
[3]CDES, OMIJ, Hôtel Burgy, 13 rue de Genève, 87065 Limoges, France

Abstract

In France, equestrian sports and leisure have been growing since the 1990s. Consequently the number of equestrian sporting events has multiplied. In this context, we wonder in what ways these events could participate in local economic development. To answer this question, we built a research program aiming at analysing economic, social and environmental impacts, in the short and long run, of different kinds of equestrian sporting events. Our first results are presented here and focus on the economic impact of small to medium size events. Using economic base theory, we studied the economic effects of six sporting equestrian events which took place in various French regions. The total economic impact of these events for the county has always been evaluated as higher than the organizational budget. Our results highlight the role of these small to medium size events for rural development of local areas and can lead to recommendations to improve economic impact of equestrian sporting events.

Keywords: economic impact, equestrian events, regional development, economic base theory

Introduction

In France, equestrian sports and leisure have been growing since the 1990s. As a result, the number of the French Equestrian Federation (FFE) members has increased from 266,000 in 1992 to almost 700,000 in 2014 (REFErences, 2014). It is today the third national sport federation in terms of members' number (behind football and tennis). Following the same trend, the number of equestrian events has multiplied. For example, 98,000 equestrian sporting contests were organized in 2014 (REFErences, 2014) compared with 17,200 in 2004 (REFErences, 2005), so their number has been multiplied by 5.7 in 10 years. These equestrian sporting events create economic impact for the rural and suburban areas where they take place. They also give rise to social utility and environmental impact. However, these consequences have never been studied or evaluated despite the challenges they emphasize, for the horse industry as well as for local areas or for the tourism sector. These factors led us to focus on this issue through the research program 'Horse and Territory'. Our goal is to build a methodology to evaluate economic, social and environmental impacts of small, medium, and large equestrian sporting events, in the short and long run. This research project started in 2011 and will run through the end of 2016. Its first step is dedicated to the economic short run impact of small to medium size events (Vial *et al.*, 2013). In this aim, we refer to base theory which is traditionally used to evaluate economic impact of mega events (that is to say big international events as world cups, Olympic games, etc.) (Barget and Gouguet, 2010a). The originality of this work is to adapt this methodology to events with two new characteristics: (1) small or medium size, which impact local areas, and (2) events of sport or leisure of open air. These characteristics have consequences for the methodology and theory. This article aims at presenting the results of a first test to assess the economic importance of small to medium size equestrian events. We focused here on six sporting equestrian events which took place in different French regions in 2012 and 2013.

The first part of this article is dedicated to the presentation of the study context in France related to the evolutions of peri-urban and rural areas on the one hand and equestrian activities on the other hand. Then, we introduce our methodology. In the third part, we present our results before discussing them in the conclusion.

Context of the study

The evolution of the countryside and equestrian activities in France: to meet the growing leisure needs of the population

For thirty years, the roles and perceptions of the French countryside have diversified. In addition to productive and historical functions (agricultural, forestry and local industry), residential and recreational functions related to the use of these areas for housing and recreation have emerged, as well as ecosystem functions (Perrier-Cornet, 2002). This transformation of lifestyles and urban-rural relationships induces a space needs for the expansion of the city and for 'nature' spaces used by urban citizens for recreational purpose (Urbain, 2002).

In the trend of growing popularity for outdoor sports and leisure, the development of the sport and leisure sector is, within the horse world, a milestone of the past twenty years (Beaumet and Rossier, 1994). It has led to an increase in the number of horses in France: their number was evaluated around 350,000 in 1995 and it increased rapidly to 570,000 in 2001 (Lemaire, 2003) and approximately one million today (REFErences, 2013). Moreover, equestrian disciplines are diversifying with the development of activities like pony riding, horse-ball, horseback hiking, reining, natural horsemanship, etc. (Digard *et al.*, 2004).

So, the evolution of the countryside and equestrian activities in France are moving towards a common goal: meeting the recreational needs of the population. However, little data is currently available about how equestrian activities, including sport and leisure, are developing and about their impact on rural and suburban areas.

A study about economic impact of equestrian sporting events

Equestrian sports and leisure create jobs and foster a local economic dynamic. Moreover, these activities take place in suburban areas as well as in rural countryside, where they generate income from the leisure spending of permanent or temporary residents. Inside the equestrian sport and leisure sector, equestrian events are one of the major sources of tourist influx, as they attract horse riders as well as people who don't have specific knowledge about horses. In the context of equestrian sport and leisure development, the number of equestrian sporting events has multiplied. Therefore, it seems reasonable to assume that they have economic consequences for the rural and suburban areas. However, very little data is currently available about these effects. Our contribution, then, is to analyse to what extent equestrian events participate in local economic development and what are their economic, social and environmental effects for areas in which they are taking place.

This issue is currently explored as part of the research program 'Horse and Territory'[1], in association with the Centre for Law and Economics of Sport (CDES) of Limoges, and the National Research Institute of Science and Technology for Environment and Agriculture (IRSTEA) of Clermont-Ferrand. Our goal is to build a methodology to evaluate the economic, social and environmental impacts of small, medium, and large size equestrian sporting events, in the short and long run. In the first step of this project, we focused on economic short run aspects of small to medium size and open air events.

To start with, in 2011, we conducted surveys during equestrian events responding to these criteria, and of various disciplines in order to understand the different public profiles and to identify the

---

[1] The research program 'Horse and Territory' began in 2006. It associates the French Institute of the Horse and Horse Riding (IFCE) and the National Institute of Agronomic Research (INRA). It is funded by the Scientific Council of the IFCE and aims at studying the economic organization of equestrian activities and their impact on regional development.

specific characteristics of these events. This work enables us to start adapting the methodology of economic impact evaluation to the specificities of equestrian events.

Then, we tested our methodology, improved it and collected data during equestrian sporting events which took place in different French regions in 2012 and 2013. They were chosen for their various disciplines, their various geographical locations and their small to medium sizes. The precise boundaries between small, medium and mega events are difficult to establish. The main criterion is that small and medium size events impact local areas and not national areas, but the other ones are difficult to define and will be discussed in the conclusion. We studied here six small to medium size events:

- The 'Equirando 2012' (Basse-Normandie region): European gathering of horse trekking riders (E 2012).
- The 'Grand Complet 2012' (Basse-Normandie region): international competition of eventing (GC 2012).
- The international jumping (4*) of Bourg-en-Bresse of 2013 (Rhône-Alpes region) (IJBB 2013).
- The regional jumping championship of Normandie of 2013 (Basse-Normandie region) (RJN 2013).
- 'Poneys sous les pommiers 2013' (Basse-Normandie region): national competition for ponies in 5 disciplines (PSP 2013).
- The national woman's horse ball championship of 2013 (Burgundy region) (HB 2013).

This article presents the results of the first step of our research program which aims at studying short run economic impact of small to medium size equestrian sporting events for local French areas. Defining the size of the event (small, medium or big) is a real question. What are the exact criteria of each category? This work will also help us to give a clear definition of these categories.

Methodology

Measuring the economic impact of sporting events raises controversies among economists about two main subjects. On one hand, we must select a suitable theoretical basis, and on the other hand, we must avoid a lot of methodological and calculation errors.

The base theory

The work presented here is based on an amended version of base theory and corresponds to an ex-post impact evaluation. The validity of this method has already been demonstrated in impact studies of mega events such as the Rugby World Cup 2007 in France (Barget and Gouguet, 2010a,b). The original contributions of our work are to adapt this method first to events of outdoor sports and second to events of small to medium size which affect local areas.

When we talk about the economic base, we generally try to determine the economic potential of an area which are most fundamental to its development (Davezies, 2008, Gouguet, 1981). In its standard version, base theory uses a dichotomous view of the regional economy: non-basic activities meet the local demand and basic activities are the source of regional economic development. Basic activities respond to an external demand: exports of goods (production base) or local consumption of goods or services by individuals whose incomes are external (recreational and residential base) (Pecqueur and Talandier, 2011). So, they enable money to come into the area thanks to an activity created by local inhabitants. Finally, basic activities create dynamism and monetary influx. Sporting events can be considered a basic activity (Bourg and Gouguet, 1998).

It seems that the base theory may be appropriate for small-scale areas to explain their economic development (Gouguet, 1981). We decided to test this hypothesis working at two geographic scales for each studied event:

- A group of 11 to 25 communities around the town where the event took place (10,000 to 73,000 inhabitants and 91 to 284 km²).
- The county (290,000 to 600,000 inhabitants and 5,700 to 8,600 km²).
- The economic system of the event highlights two main sources of injections:
- Consumer spending comes from spectators, participants and accompanying persons who came with participants.
- Spending related to the organization of the event.

Note: Investment spending is not taken into account because small events do not directly involve constructions or renovations.

Methodological improvements

The economic impact of mega sporting events is frequently overvalued. The reason is often mistakes in economic calculations related to an insufficiently rigorous methodology (Jeanrenaud, 2000; Preuss, 2006). We tried to avoid the most common:

- Not taking into account the substitution effect. First of all it concerns spending coming from local agents. If the event hadn't been organized, local officials would have certainly spent money on another purpose in the area. Only spending from agents coming from outside the area is considered to determine the net injection. Similarly, among the organization spending, it is necessary to take into account the origins of funds that have been used (they must be external to the study field to be taken into account), and the purpose of the expenditure (they must benefit domestic agents inside the study field to be taken into account).
- Not taking into account the reason for the visit. We were able to identify the reasons why visitors came (the event, tourism, visiting family, professional reasons...). If someone came only for the event, we took into account his entire spending. If he didn't come for the event, his spending is not taken into account. Finally, if the event was not the only reason for coming, we took into account a fraction of the expenditures (for example, if a respondent indicated two reasons for coming, we included 50% of the expenditures). We did not try to assess the temporal shift in consumer spending. This applies to spectators who would have delayed their trip (sooner or later) to come during the event. Indeed, we consider here that this is not relevant for small to medium size events, by contrast with mega events. Indeed, spectator probably often delay their trip to be able to attend mega events as they are very important and famous but we suppose that this behaviour is very marginal for small to medium size events that are less popular.
- Not taking into account the leakages outside the study zone, that is to say the money that leaves the area because of the event needs (purchase of imported products or services). This often leads to an overestimation of the impact when the gross injection is taken into account instead of the added value which allows removing the leakages due to intermediate consumptions or imported products. It is also necessary to distinguish wages from purchases of goods and services. All wages are considered as additional income paid locally. For purchases of goods and services, we must subtract all leakages and therefore use the added value rate. Unlike in the Anglo-Saxon countries, regional statistics in France didn't identify the flows of goods and services between regions. Under these conditions, we used bibliographical reviews about different studies on the subject conducted in France to determine the added value. An average, we estimated it at 50% for groups of communities at the local level and at 60% for counties (Barget and Gouguet, 2010a,b).
- Double counting for injections (for example by summing the commitment costs of participants and organization expenses funded by those revenues, which means counting the same amount of money twice).

- Finally, we ignore the crowding-out effects on consumption. Indeed, we assume that, given the small size of the events, their organizations will not have discouraged potential visitors to come because of the fear of local road congestion or of nuisances created by the event. We make this hypothesis because in the case of small to medium size events, nuisances for local inhabitants are insignificant, so the number of tourist who could have cancel their venue because of the event is considered as really negligible.

All these deficiencies generally lead to an overestimation of the economic impact. To address these shortcomings, which are related to a deficiency in data collection and analysis, we attempted to collect better quality data in the field.

Collecting field data

In addition to the regional information available (INSEE -National Institute for Statistics and Economic Studies-), we collected information specific to the event in two ways:
- Face-to-face interviews (on the basis of a closed questionnaire) among participants, accompanying persons, spectators, employees, volunteers, sellers and restaurateurs (Table 1).
- Semi-structured interviews with the organizers, that is to say open discussions on the basis of a questionnaire, and collection of the accounting of the event.

Table 1. Number of interviewed people (processable) and total number of people for each studied event.

	E 2012[1]		GC 2012[2]		IJBB 2013[3]		RJN 2013[4]		PSP 2013[5]		HB 2013[6]	
	Interviewed people	Total	Interviewed people	Total	Interviewed people	Total	Interviewed people	Total	Interviewed people	Total	Interviewed people	Total
Participants	40	1,184	90	250	54	596	161	1,169	198	2,400	59	1,042
Accompanying persons	25	500	301	15,000			21	28	62	210	151	1,033
Spectators					316	9,800						
Employees	16	40	11	11	28	100	4	8	24	60	10	13
Volunteers												
Sellers	9	9	0	100	22	55	14	30	11	28	6	19
Restaurateurs												
Total	90	1,733	402	15,361	420	10,551	200	1,235	295	2,698	226	2,107

[1] Equirando 2012: European gathering of horse trekking riders (Basse-Normandie region).

[2] The 'Grand Complet 2012': international competition of eventing (Basse-Normandie region).

[3] The international jumping of Bourg-en-Bresse of 2013 (Rhône-Alpes region).

[4] The regional jumping championship of Normandie of 2013 (Basse-Normandie region).

[5] 'Poneys sous les pommiers 2013': national competition for ponies in 5 disciplines (Basse-Normandie region).

[6] The national woman's horse ball championship of 2013 (Burgundy region).

Moreover, it was necessary to estimate the numbers of people for each of the above categories. This was accomplished in the following ways:

- The number of participants was provided by the organizers, as registration was mandatory and fees charged.
- Similarly, the number of employees, volunteers, sellers and restaurateurs was known by the organizers.
- The average number of accompanying persons was estimated from the participant questionnaire, which included a question about the number of accompanying persons.
- One of the events studied charged fees for spectators, and its access was closed. Hence, it was possible to obtain the number of spectators from ticket information. For other events, the number of spectators was assessed through different methodologies, including counting vehicles on spectator parking (noting the number of plates at different times during the event or pasting a sticker on each vehicle) or bracelet distribution during the event.

The multiplier calculation

The total economic impact corresponds to the net injection multiplied by a multiplier. This ratio enables to take into account the multiplier effect of the money injected on the area, that is to say the 're-use' of this money on several wages after the event. Multipliers are estimated from a meta-analysis (Vollet and Bousset, 2002) using the model of the economic base. We calculated a multiplier for each kind of people category which gives rise to an injection (spectators, accompanying persons, participants...), using their propensity to spend locally. We therefore obtained an aggregated multiplier of basic type (Bourg and Gouguet, 1998). The basic formula to calculate the multiplier K is the Wilson formula (Wilson, 1977):

$$K = (1 - m1 + m2) / (1 - m2)$$

where $m1$ = propensity to spend in the first wave and $m2$ = propensity to spend in the other waves.

Table 2 shows the results of the multiplier estimation. Each calculated multiplier is specific to each geographic level and to each public category. However, we only present here an average multiplier for each geographic level and for each event studied (example: the average multiplier for RJN 2013 for the county is a mean of the multiplier of spectators for RJN 2013 for the county, the multiplier of participants and accompanying persons for RJN 2013 for the county, the multiplier linked to the organizational budget for RJN 2013 for the county, the multiplier of employees and volunteers for RJN 2013 for the county, and the multiplier of sellers and restaurateurs for RJN 2013 for the county).

Table 2. Multipliers[1].

	Local group of communities	County
RJN 2013	2.06	2.20
HB 2013	1.92	1.93
PSP 2013	1.30	1.31
E 2012	1.86	1.66
GC 2012	1.90	1.45
IJBB 2013	1.75	1.38

[1] See Table 1 for an explanation of the abbreviations.

Results

Injections related to organizational expenditure

Organizational budgets are presented in Table 3. Revenues have four sources: participant inscription spending, grants, sponsors, other sources of income (rental and sales). A detailed analysis of each recipe allowed us to identify if it comes from inside or outside the studied areas. This enabled us to calculate the percentage of the budget that comes from outside each study field (in other words, the percentage that allows us to assess the share of spending that comes from an external source of funding). We also calculated the amount of money spent inside each study field. By multiplying these two figures, we obtained the gross injection. Finally, the added value rate enabled us to calculate the net injection. The results are presented in Table 3.

This analysis of the organizational budget is necessary to evaluate the impact of the event. Nevertheless, it also leads to some recommendations to improve the economic impact of the event. First, we can see that the percentage of the organizational budget coming from outside the study zone varies from 43 to 99%. If possible, organizers should try to increase this percentage in order to improve the economic impact of the event; in other words, they should look for sponsors and grants coming from outside their area. Second, the percentage of the organizational budget spent inside the

Table 3. Analysis of the organizational budget of each event.

Event	Organizational budget (OB) (€)	Study field	Percentage of the OB coming from outside the study field	Amount of the OB spent inside the study field (€)	Gross injection (€)	Net injection (€)
RJN 2013	23,000	Group of local communities	96%	2,115	2,031 (including 696€ of wages)	1,364
		County	52%	9,484	4,932 (including 1,052€ of wages)	3,380
HB 2013	50,000	Group of local communities	99%	10,264	10,162	6,582
		County	99%	18,101	17,920	11,952
PSP 2013	92,000	Group of local communities	92%	25,843	23,776 (including 850€ of wages)	12,313
		County	81%	32,930	26,674 (including 5,790€ of wages)	18,321
E 2012	283,000	Group of local communities	99%	7,390	7,317	3,659
		County	92%	48,005	44,165	26,499
GC 2012	373,000	Group of local communities	78%	41,173	32,115	16,057
		County	65%	67,218	43,692	26,215
IJBB 2013	500,000	Group of local communities	43%	231,944	99,736	49,868
		County	44%	240,893	105,993 (including 940€ of wages)	65,535

study zone varies from 9 to 47%. In order to improve the impact of the event, the organizers have to spend as much money as possible locally. They should foster relationships and contracts with local suppliers and partners.

Injections related to the different public categories

Among the total interviewed people, most of them come only for the event. Of course, the percentage is higher for participants and accompanying persons (95 to 58% according to the event with a mean at 81%) than for spectators (95 to 45% with a mean at 67%). Employees, volunteers, sellers and restaurateurs come almost only for the event.

People's geographic origin depends on the event but there is always a majority of them who come from outside the two study zones (almost all the participants and between half and all of other public categories comes from outside the group of local communities). Less people come from outside the county (between 44 and 97% of the participants and between 20 and 80% for other public categories).

The average expense linked to the event depends on the kind of event, its length, the public category and their origin. On average, participants spend between 124 and 619€ at the event, which is between twice and 15 times more than spectators.

For each public category, we also needed the number of people attending the event. As we explained in the methodological section, the main problem is accounting for spectators. For open events, the number of spectators was assessed through the different methodologies we wanted to test:
- counting vehicles on spectator parking, noting the number of the plates at different times during the event;
- counting vehicles on spectator parking by pasting a sticker on each vehicle (each vehicle receives a sticker and has to display it during the entire event so the number of distributed stickers is equivalent to the number of vehicles);
- bracelet distribution during the event (each spectator receives a bracelet and has to wear it during the entire event so the number of distributed bracelets is equivalent to the number of spectators).

Each of these methodologies was complemented by specific questions in the questionnaires (for example the questionnaires enabled us to know the mean number of spectators per vehicle and consequently to access the total number of spectators in multiplying with the number of spectator vehicles). Every method enabled us to obtain estimations of the spectator number. Nevertheless, one methodology can be more appropriate than another according to different characteristics. When parking areas are relatively well defined, vehicle counting on spectator parking works well and is cheaper. Pasting a sticker on each vehicle seems better than noting the number of the plates because it is less time consuming; however, it requires good meteorological conditions, since heavy rain can remove stickers. However, when parking areas are diffuse, a bracelet distribution is more accurate but more expensive.

Using the geographic origin of interviewed people, and their reasons for coming, the total numbers of people attending the event, the amounts of money spent in the different expense categories and in the different study zones, we calculated the gross injection. Then we deduced the net injection due to the added value rate. What seems interesting is to compare the proportion of the net injection coming from each public category and from the organization (Figure 1).

Figure 1 highlights the various weights of the different injection sources. In every case, sellers, restaurateurs, volunteers and employees represent a small part of the injection. Consequently, the value in investigating them may be small. The organization represents 10 to 20% of the injection, and we have seen previously that it can still be improved by organizers by increasing the percentage

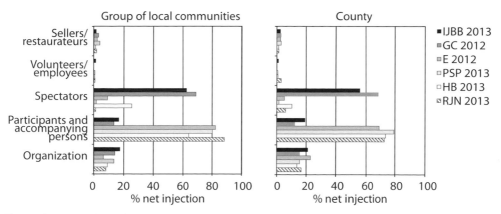

Figure 1. For each study zone, percentage of the net injection coming from each public category for each studied event.

of the organizational budget coming from outside the local area and spending this money inside the local area. Participants, accompanying persons and spectators represent the major sources of injection. On these graphs, we notice an interesting difference between small and medium events. In small events, the monetary injection is mainly due to participants and accompanying persons (approximately 80% of the injection), as these events don't attract people coming from far away. Whereas in medium events, however, the monetary injection is more due to spectators (60% of the injection), as these events attract people who come from other counties. It seems reasonable to assume that this difference would be more important for mega events which attract people from all over the world.

These results also suggest recommendations to improve net injections. For small events, the monetary injection of spectators could be improved by attracting more spectators and encouraging them to spend more money and to stay longer. In this aim, organizers could improve the communication around the event through advertising in the press, tourist offices, tourist places, internet, etc. They could also encourage the participation of people who don't know anything about horse riding by adapting the event to those people (activities like short pony rides for children, introductions to different horse riding disciplines, adaptation of the speaker's comments, etc.). For medium events, the monetary injection of participants and accompanying persons could be improved by encouraging them to spend more money and to stay longer. In order to achieve this goal, organizers could encourage them to stay in hotels instead of sleeping in their trucks or tents or to eat out instead of bringing their food, to participate in tourist activities, and so on. This can be done through advertising for various housing and local restaurants via the event website and through flyers (with discounts) distributed to participants. Organizers could also propose activities, such as touristic visit to accompanying persons who have more time than participants (for example through a partnership with the tourist office).

The economic impact evaluation

Finally, our analysis enabled us to calculate the economic impact of each event. After evaluating the net injection, the use of multipliers enabled us to add the induced effects in order to obtain the total impact of each event at each geographic level (Table 4).

As we can see in Table 4 and Figure 2, the total economic impact is higher than the organizational budget in all the cases except the two events studied in 2012. We suppose that this can be due to the methodology that was not precise enough and finished in 2012. These two events (E 2012 and

Table 4. Evaluation of the total economic impact of each event.

Event	Study field	Gross injection (€)	Net injection (€)	Multiplier	Total economic impact (€)
RJN 2013	Group of local communities	33,000	17,000	2.06	35,000
	County	33,000	20,000	2.20	44,000
HB 2013	Group of local communities	141,000	72,000	1.92	138,000
	County	143,000	87,000	1.93	168,000
PSP 2013	Group of local communities	180,000	91,000	1.30	118,000
	County	192,000	204,000	1.31	268,000
E 2012	Group of local communities	114,000	57,000	1.86	106,000
	County	193,000	115,000	1.66	191,000
GC 2012	Group of local communities	226,000	113,000	1.90	215,000
	County	277,000	146,000	1.45	241,000
IJBB 2013	Group of local communities	566,000	283,000	1.75	496,000
	County	512,000	388,000	1.38	534,000

GC 2012) were our first case studies; they served as pilot studies. So, the validity of their results is limited and we feel that the economic impact evaluation is underestimated for several reasons:
• the number of people interviewed is too small;
• the counting of spectators and accompanying persons was incomplete and we thus counted a number of each people category inferior to the reality;
• the impact of sellers, restaurateurs, employees and volunteers was not taken into account.

All these limits were improved in the next studies.

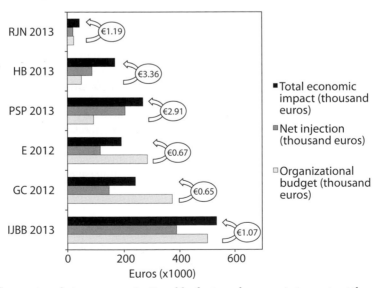

Figure 2. Comparison between organizational budgets and economic impacts at the county level.

If we compare the total economic impact and the organizational budget, we can see the amount of money which is generated by the event. To do that, we divided the total economic impact by the organizational budget. At the county level, if we don't take into account the events studied in 2012 for previous reasons, our results show that small events (RJN 2013, HB 2013 and PSP 2013) would be more cost-effective than big events (IJBB 2013): for one euro invested in the organization of RJN 2013, HB 2013 or PSP 2013, the economic impact for the county reaches 1.91 to 3.36 euros whereas it only reaches 1.07 euros for IJBB 2014 (Figure 2).

Conclusion

Our final aim is to build a methodology to evaluate economic, social and environmental impacts of equestrian sporting events. The first step of our research program focuses on short run economic issues. For this objective, we used a methodology that refers to economic base theory. This paper presents the results of two rounds of tests on six equestrian events in France. Our findings enabled us to have a better knowledge of the different public categories and showed how important their economic impact is for local areas. But far more important, this work enabled and will still enable us to improve our methodology.

Beyond all the methodological precautions taken to avoid the most common mistakes which often leads to an overestimation of the impact (not taking into account the substitution effect, the crowding out effect, the reason for the visit, the leakages outside the study zone, and double counting for injections), the originality of this work is to think about the adaptation of this method to events with two characteristics: (1) small to medium size events, that is to say events which impact local areas; and (2) events of sport or leisure of open air. The first characteristic –small to medium size events- required adaptation of multiplier calculations, geographic level of study zones, and result analysis. The second one – events of sport of open air – required us to develop a methodology to count spectators. We will now discuss our methodology.

As we said before, the validity of the tests for events held in 2012 is limited for several methodological reasons; however, they led to methodological improvements for the tests done in 2013. These two experimental years enabled us to build a better methodology which is currently being tested on other equestrian events. Some new methodological improvements can be recommended. For example, the questionnaires have to be short to maximize response rate and quality of responses (10 minutes maximum and well-constructed). We think that it is necessary to interview spectators, participants and accompanying persons, but not the sellers, restaurateurs, employees and volunteers whose economic impact is low comparing to the other previous public categories. In an open area, the counting of spectators can be done in two ways according to the event characteristics. When parking areas are relatively well defined, we recommend vehicle counting on spectator parking through pasting a sticker on each vehicle. However, when parking areas are diffuse, a bracelet distribution is more appropriate even though it is more expensive.

Like other leisure activities or sports, equestrian activities are a source of direct economic benefits for local areas. Among them, small or medium equestrian events impact local areas where they take place and can be considered as basic activities for these areas. But what is a small or a medium event? In this work, we developed a gradient with the studied events according to their organizational budget. Our results highlight a distinction between two kinds of events:
- ones whose economic impact is mainly due to participants and whose organizational budgets are smaller than three hundred thousand euros;
- ones whose economic impact is mainly due to spectators and whose organizational budgets are higher than three hundred thousand euros.

So, it seems that this organizational budget criterion could be used to distinguish small from medium size events. We suppose that mega events will look like medium events, as their economic impact is mainly due to spectators as well, but they can be defined as international and well-known events (such as the Olympic Games and the World Equestrian Games). However, this classification may need to be improved again, using other criteria that could depend on the discipline, the area, and so on.

Finally, our first tests highlight the importance of studying the impact not only of mega events but also of small or medium local events, as they impact local areas. Moreover, it shows that the methodology can be adapted to events of open air. Our aim is now to take into account not only the short term economic impact of equestrian sporting events but also their externalities and long term role in local economic development, as well as their social and environmental impacts. New tests on these themes are currently underway.

References

Barget, E. and J.J. Gouguet, 2010a. De l'évaluation des grands événements sportifs. La Coupe du Monde de Rugby 2007 en France. PULIM, Limoges, France, 386 pp.

Barget, E. and J.J. Gouguet, 2010b. Hosting mega-sporting events: which decision-making rule? International Journal of Sport Finance 5(2): 141-161.

Beaumet, M. and E. Rossier, 1994. Les loisirs à cheval. Etude réalisée pour l'Agence Française de l'Ingénierie Touristique et la Direction de l'Espace Rural et de la Forêt du Ministère de l'Agriculture et de la Pêche, 74 pp.

Bourg, J.F. and J.J. Gouguet, 1998. Analyse économique du sport. Presses Universitaires de France, Paris, France, 380 pp.

Davezies, L., 2008. La République et ses territoires: la circulation invisible des richesses. Editions du Seuil, Paris, France, 96 pp.

Digard, J.P., L. Ould Ferhat, C. Tourre-Malen, A. Caporal and N. Vialles, 2004. Cultures équestres en crise: professionnels et usagers du cheval face au changement. Les Haras Nationaux, Ivry-sur-Seine, France, 39 pp.

Gouguet, J.J., 1981. Pour une réhabilitation de la théorie de la base. Revue d'Economie Régionale et Urbaine 1: 63-83.

Jeanrenaud, C., 2000. The economic impact of sport events. Editions Centre International d'Etude du Sport (CIES), University of Neuchatel, Neuchatel, Switzerland.

Lemaire, S., 2003. Économie et avenir de la filière chevaline. INRA Productions Animales 16: 357-364.

Pecqueur, B. and M. Talandier, 2011. Les espaces de développement résidentiel et touristique: etat des lieux et problématiques. In: DATAR Territoires 2040: 120-138.

Perrier-Cornet, P., 2002. Repenser les campagnes. Éditions de l'Aube, La tour d'Aigues, France, 280 pp.

Preuss, H., 2006. Special issue: impact and evaluation of major sporting events. European Sport Management Quarterly 6(4): 313-415.

REFErences (Réseau Economique de la Filière Equine), 2005. Annuaire ECUS 2005: tableau économique, statistique et graphique de cheval en France. Institut Français du Cheval et de l'Equitation, Saumur, France, 63 pp.

REFErences (Réseau Economique de la Filière Equine), 2014. Annuaire ECUS 2014: tableau économique, statistique et graphique de cheval en France. Institut Français du Cheval et de l'Equitation, Saumur, France, 63 pp.

REFErences (Réseau Economique de la Filière Equine), 2013. Annuaire ECUS 2013: tableau économique, statistique et graphique de cheval en France. Données 2012-2013, Institut Français du Cheval et de l'Equitation, Saumur, France, 63 pp.

Urbain, J.D., 2002. Paradis verts. Désirs de campagne et passions résidentielles. Éditions Payot, Paris, France, 387 pp.

Vial, C., E. Barget and J.J. Gouguet, 2013. Economic impact and social utility of equestrian events, examples from France. In: Book of abstracts of the 64th Annual Meeting of the European Federation of Animal Science. Wageningen Academic Publishers, Wageningen, the Netherlands, p. 253.

Vollet, D. and J.P. Bousset, 2002. Use of meta-analysis for the comparison transfer of economics base multipliers. Regional Studies 36(5): 481-494.

Wilson, J.H., 1977. Impact analysis and multiplier specification. Growth and Change 8(3): 42-46.

2. Input-output analysis of the Swedish and Norwegian horse sectors: modelling the socio-economic impacts of equine activities

G. Lindberg[1], A. Spissøy[2] and Y. Surry[3]*
[1]Nordregio, Box 1658, 111 86 Stockholm, Sweden
[2]County Governor of Hordaland, Box 7310, 5020 Bergen, Norway
[3]Swedish University of Agricultural Sciences, Box 7013, 750 07 Uppsala, Sweden; yves.surry@slu.se

Abstract

The economic impacts of equine activities in Norway and Sweden were analysed and differences between effects at national and regional levels were examined in detail. An input-output (IO) model was used to examine linkages and impacts. Unfortunately, horse-related information is currently either lacking or distributed across different sectors of the economy in national IO accounts. We therefore examined earlier attempts to separate a horse sector in IO models and devised a simple method for disaggregating such accounts based on different sources of data including surveys, interviews and disaggregated sector data. We developed IO coefficients for horse-related activities in Norway and Sweden and used these coefficients to derive multipliers showing the impacts on the economy of expanding different horse activities. We also examined the regional structure of the multipliers. For Sweden, we found the highest multipliers for riding schools (3.19) and breeders (2.90). The reason for the greater magnitude of these multipliers may be that such enterprises use most of their revenue for purchasing inputs and spend it in the supply chains, and hence do not make much profit. Activities needing to allocate some revenue to returns on capital and wages, e.g. boarding enterprises (2.86) and professional trainers (2.61), had somewhat lower multipliers.

Keywords: equine activities, input-output analysis, Norway, Sweden

Introduction

The horse industries in Norway and Sweden have some interesting similarities in terms of economic and social importance. The betting industry constitutes an important income source for the sector in both countries. The industries are organised in a relatively similar way, where a predetermined share of the turnover from gambling is returned to the sector (ATG, 2011; Gren, 2006). However, some differences are apparent. Different structures of the horse industry may be explained by differing historical, agricultural and financial structures. At present, more horses are kept for leisure, riding and other recreation purposes in Sweden, where there are approximately 30 horses per thousand individuals, while the corresponding number in Norway is 24 (Liljenstolpe, 2009).

The explanation for this may be complex. For instance, in Sweden there is a long knighthood tradition, whereas this is non-existent in Norway apart from a limited military tradition. Rules of taxation may also have had some impact on the structures, e.g. it has become more common for Norwegian horse owners to register their horses in Sweden due to the more favourable value added tax (VAT) rules for horses in private companies in Sweden (Econ Pöyry, 2008). Another important difference is that Sweden is a member of the European Union (EU), and the equine industry has potential for future growth and diversification within the EU Rural Development Programmes, with payment schemes intended to enhance horse grazing and production of hay in order to preserve open landscape and with investment support (Liljenstolpe, 2009). Moreover, the physical conditions for horse keeping are arguably more favourable in Sweden, as the area of land required to keep a warmblood horse, assuming that all feed is produced domestically, is 1.18 ha for Sweden and 1.37 ha for Norway (Liljenstolpe, 2009).

It is important for the industry and for policymakers to understand the economic and social impacts of potential growth or structural changes in the equine industry. This includes understanding the interactions with both suppliers and consumers of intermediate and final products. An input-output (IO) model is a suitable tool for analysing such linkages and impacts (Leontief, 1966), but for the horse sector input-output information is currently lacking in national IO accounts and in distributed accounts for different sectors in the economy. For instance, agriculture is usually aggregated into a single sector in national accounts, whereas riding schools are part of recreational activities. The present analysis sought to examine how the horse sector can be studied from an economic linkage perspective and to provide examples of this for two Nordic countries. The information developed conceptually and empirically in this study can be useful for scenario analysis and for policy discussions about rural development and horse activities.

The aim of the study was to analyse the horse sector from a national and regional perspective using an IO modelling approach. IO modelling is a frequently used tool for analysing the economic importance of different sectors and for analysing the impacts of actual or proposed changes (Hughes, 2003). IO coefficients are basically average values describing the quantities of inputs used in production of a commodity or industry. We collected such data using surveys, databases and interviews, as described below. IO coefficients can be used to derive multipliers showing the direct, indirect and induced impact that each sector has on the economy when it experiences a positive or negative shock. Put less formally, multipliers are a way to look at the 'rings on the water' when something impacts on the economic landscape or surface. A multiplier shows how much the economy (and what sectors) would grow as a result of an increase in final demand (and hence production) of a sector. As the sector itself expands, it will require its interacting sectors to grow as well, to supply physical and non-physical inputs, and this effect will work its way through the economy. This is one way to investigate what kind of impact the horse-related sector can have on the economy, both today and if it were to grow due to expansion of riding schools, horse tourism or gambling, and consequently horse breeding.

We derived multipliers based on the IO coefficients we developed for the horse sector in Sweden and Norway and examined the regional structure of such multipliers. National tables are usually regionalised using employment data. As such data do not exist for all horse sub-sectors, we used the regional horse numbers, together with employment data, to regionalise our structural data to county level in both Sweden and Norway.

IO analysis as a method to study the horse sector

Input-output accounting is a method for analysing the economic interactions between sectors in an economy and as such it provides a practical 'database', or matrix, of economic interactions between sectors, households and other institutions. The purpose of IO analysis is usually to derive multipliers measuring the importance of direct and indirect linkages in the economy.

The IO table is an accounting identity describing in a static fashion the linkages within an economy at a specific point in time[1]. The table records the various financial interdependencies between sectors in the economy and their consumption of intermediate goods and services, distinguishing domestic and imported goods and services. It also reports the final consumption expenditure, including exports, by households, government and other institutions. For each sector it is possible to distinguish payments to households, capital costs, taxes and subsidies and other final payments. The IO model assumes that the economy of a country or region can be divided into a specific number of well-defined sectors. The sectors can be larger or smaller depending on the level of aggregation. Furthermore, the IO table is a fixed price equilibrium model utilising the Leontief production function of fixed proportions.

---

[1] Recent classic texts on IO modeling are Miller and Blair (2009) and Ten Raa (2005), while agricultural and rural development applications of IO analysis can be found in Midmore (1991) and Midmore and Harrison-Mayfield (1996).

However, important economic assumptions in this model should be acknowledged and remembered in analysis of an economic system (Miller and Blair, 2009).

First assumption: Each sector produces only one good. This assumption implies that the flows of goods associated with firms producing two or more goods must be allocated to two or more production lines. Each production line can be considered an individual firm that interacts with other individual firms. Thus, IO modelling does not take the possible benefits of joint production into consideration.

Second assumption: Fixed proportions/fixed coefficients of production. This implies constant returns to scale, which means that the use of inputs is proportional to the outputs of the firm, regardless of size. Put another way, if inputs are doubled, so are outputs. Even though constant returns to scale can be argued to simplify description of reality, micro-economic theory shows that it is the long-term equilibrium for firms operating in a perfectly competitive market.

Third assumption: No lack of capacity. Through this assumption, the economy is assumed to immediately satisfy the demand for extra production inputs. For instance, if more labour is demanded, it is immediately supplied.

IO tables report transactions of an economy in terms of either industries or commodities. That is, the underlying table accounts show either industries or commodities as activities or sectors of the economy. The Swedish IO table, for instance, is built in a commodity-by-commodity framework. The extended 'make and use' format, which is often the basis for national IO tables, shows how industries use commodities and what mix of commodities each industry supplies. This can often result in a rectangular table system, as there may be more or less commodities than industries, so in order to analyse the inter-industry structure of an economy, it is converted to a square IO table.

In the square IO system, technical coefficients show each industry's use of intermediate inputs regardless of their country of origin. Trade coefficients show each industry's use of intermediate inputs produced domestically[2]. Technical or trade coefficients are calculated for the transaction table of intermediate inputs as the ratio of input (total or domestic) to output in each sector. Denoting the technical coefficient for the use of sector i's product in the production of sector j's output as a_{ij}, then $a_{ij} = Z_{ij} / X_j$, where Z_{ij} is the economic flow from sector i to j, and X_j is the total output value of sector j. Gathering technical or trade coefficients in a matrix A enables the vector of outputs, X, to be expressed as $X = AX + Y$, where Y is the vector of exogenous final demand. The equation $\Delta X = (I - A)^{-1} \Delta Y$ shows how a change in final demand (ΔY) affects total output (ΔX) through backward linkages in the economy in the form of multipliers $(I - A)^{-1}$.

In the model outlined above, final demand (Y) is exogenous and not part of the interrelated production system. The final demand consists of purchases by households, government, private investors and exports. For households, treating this component of the economic system as an exogenous element in impact analysis is questionable. Households earn income from the other sectors in the economy in the form of wages, which are used to purchase goods from producing sectors. Therefore, one could argue that labour income and consumption by households should be 'added' to the X matrix. This would involve adding one more row and one more column to X, hence making wages and household final demand endogenous to the input-output model. The additional row would show how the labour earnings, i.e. wages and other fringe benefits, are distributed. The consumption column would show what goods and services households purchase. This procedure is called *closing* the IO model with

---

[2] Trade coefficients are the relevant measure for analysing impacts in most instances for a region or nation, since the technological coefficients disregard leakages of impacts abroad. Multipliers were calculated here using trade coefficients.

respect to households. A model which disregards these induced effects of household earnings and consumption is denoted *open*.[3]

Previous studies of the horse sector using input-output methods

Table 1 lists examples of economic impact studies using IO models of the equine activities conducted in the past fifteen years in the United States, Canada and a couple of European countries including Great Britain and Sweden. The last two right-hand side columns of this table give estimates of the output multipliers based on the use of an open or closed IO table. As explained earlier, an open output multiplier will only record initial direct and indirect output effects, while a closed output multiplier

Table 1. Snapshot of past economic impact studies of equine activities in North America and Europe using input-output models.

References	Country, area	Equine sectors	Output multipliers	
			Open	Closed
	Canada, Provinces			
ERL (2001)	Alberta	Horse racing and breeding		2.21
Canmac (2008)	Nova Scotia	Harness racing industry		2.15
ERL (2011)	Ontario	Horse racing and breeding		2.20
	Europe, Countries			
Deloitte (2013)	United Kingdom	British horse racing industry		2.55
Andersson and	Sweden	Thoroughbred and harness racing	1.80	3.20
Johansson (2004)		Equestrian sports	2.20	3.50
		Horse tourism	2.20	3.60
		Horse breeding	1.42	3.90
		Gambling services	1.48	1.80
		Commodities and services	1.50	2.80
		Other horse activities	1.76	2.90
		Total equine sector	1.50	2.80
	United States, States			
Beattie *et al.* (2001)	Arizona	Pleasure horses		1.60
		Horseracing		1.90
		Horse shows		1.91
		Other		1.89
		Total equine sector		1.67
University of Kentucky (2013)	Kentucky	Total equine sector		1.68
Menard *et al.* (2010)	Tennessee	Total equine sector	1.28	1.95
Hughes *et al.* (2011)	West Virginia	Pleasure equine owners and non-racing equine businesses		1.70
		State horse race industries		1.42
		Total equine sector		1.53

---

[3] These open and closed multipliers are sometimes denoted type I and type II multipliers. However, with type I and II multipliers the impact is divided by the direct effect. For output multipliers this is unity and the multipliers are the same as the open and closed, but for income and employment multipliers types I and II are not necessarily the same as the open and closed. See for instance Miller and Blair (2009) for a discussion of differences between multiplier types.

includes those effects and also induced effects resulting from endogenous household expenditure. As a result, closed output multipliers are always greater in magnitude than their corresponding open counterparts.

Not surprisingly, all open multiplier estimates are plausible with values ranging from 1.2 to 2.2, which fall within the expected range of similar multiplier estimates found for other agricultural activities (OECD, 2009). On the other hand, closed output multiplier estimates are higher than 1.5 and some for horse-related activities are even close to 4 (horse breeding in Sweden). Again, the magnitude of the closed output multipliers is in line with those reported elsewhere for other agricultural activities.

Another issue that needs to be stressed is that output multipliers tend to be smaller in size for local or regional economies than those estimated at the national level (Lindberg, 2011). This is not surprising, because regional or local economies tend to be more open, with more trade and hence leakages to the rest of the world (as well as the country) than a national economy. This pattern can be observed in the studies reported in Table 1, since the closed multipliers obtained for Sweden and Great Britain are larger in size than those obtained for Canadian provinces or US states.

Finally, it is important to observe that a significant number of economic impact studies on equine sectors use established IO tables without considering horse-related activities as separate activities. What they do instead is to consider the initial expenditures generated by equine activities and then determine the resulting multiplier effects associated with these initial expenditures. This results in overall induced impacts that can be compared with appropriate economic aggregates for the regional or national economy. This approach is used in studies reported in Table 1 for Great Britain (Deloitte, 2013), Nova Scotia (Canmac, 2008) and several US states. In the latter case, it is worth mentioning that the IO tables used for the US state equine sectors are derived from a regionalisation of the US IMPLAN IO table[4]. On the other hand, the economic impact studies of horse-related activities conducted by ERL (2001) in Canada and by Andersson and Johansson (2004) in Sweden are based on a representation of explicit horse-related activities in the IO table used.

Data collection and construction of the horse sub-sectors in an input-output table

To analyse specifically the various sectors in the economy that are related to the horse industry, it is necessary to distinguish them in an IO table. Furthermore, to perform this analysis at the regional level, it is necessary to scale down the national IO table using appropriate regionalisation procedures. At present, horse activities are spread throughout the Swedish and Norwegian IO tables in different sectors (industries). For instance, those flows relating to horses on farms are part of the transactions associated with the agriculture sector. To conduct impact analysis of the horse activities at the regional level, we disaggregated these sub-sectors from their current aggregation/position in the IO table and modelled them explicitly in relation to the rest of the regional economy. This amounted to 'breaking' these economic transactions away from the standard IO table, recording them in separate rows and columns and subtracting them from the original IO table in order to avoid double counting. Since some aspects of the horse industry may have been overlooked or ignored by national statistical offices while building the national table, this procedure also involved adding such flows to the regional or national IO tables.

The sectors included in this study are obviously not the only possible sub-sectors of the horse industry. However, they are the sectors considered most important from a rural development perspective and for which it was feasible to collect information using surveys and experts in the two regions. For

---

[4] In the United States many IO studies are conducted by the IMPLAN project (www.implan.com), which enables disaggregated and regionalised analysis using software-based IO models. For more details, see Lindall and Olson (2008).

instance, horse-related gambling is a large national/global activity for which it is difficult to elicit the economic linkages and which is difficult to break out of the national sector in which it is included. Although we did not pursue this line of thought, there might also be ethical considerations in building a demand-side scenario which is framed around increases in gambling to stimulate rural development in horse-related activities. Some sectors were excluded here since they can already be modelled in the standard IO model, e.g. veterinary medicine, because they are limited in size, e.g. horse-related media, or because they are marginal in the Nordic countries, e.g. meat. The sectors studied and the way in which data were collected in this study are described in Table 2.

In linking the data to the IO table, we should bear in mind two important aspects. First, the transactions recorded in the data are evaluated at purchaser prices. In the IO table, the trade and transport margins of all intermediate interactions are recorded in separate accounts. That is, only the 'base price' of the interactions is recorded in the column and row of the transaction. This is in order to model explicitly the trade and transport sectors of the economy. Furthermore, when dealing with a domestic table, a fraction of total intermediate use must be allocated to the imports account, so that the national economy can be properly represented. We addressed both of these aspects of the classification by assuming similar structures as in the agricultural accounts, i.e. we let the horse enterprises behave like agricultural firms in the import shares and transport costs of the products that horse firms actually purchase. For each commodity traded in the IO account, there is a record in the initial table giving the level of the trade and transport margins and there is an import table showing the level of imports of all commodities in each sector. These margins and shares were applied here to all transactions made by horse firms. This information is available for both the Swedish and Norwegian IO tables.

Regionalisation

Input-output tables are frequently used to study regional economies in an attempt to better understand economic structures or predict effects of sectoral or regional policies. To study regional economies, methods have been developed to 'regionalise' national tables. The reason for this is the high cost associated with producing survey-based regional tables. In both Sweden and Norway, the national disaggregated table has been regionalised using non-survey approaches. Non-survey regionalisation was initially conducted by some simple location quotient (SLQ) or cross-industry location quotient (CILQ), where sectors that are not large enough to support the regional demand by other sectors are corrected (row wise for SLQ and cell by cell for the CILQ). Location quotients are usually based on employment figures for the region vis-à-vis the nation but other data such as value added, output and wages have sometimes been used. To take both relative industrial size and regional size into consideration, a refined location quotient procedure has been proposed by Flegg *et al.* (1995) and Flegg and Webber (2000). This procedure, which is known as the FLQ formula, was used in this study to regionalise the IO tables to 21 regions in Sweden and one region in Norway.

Table 2. Equine activities (sectors) studied and the data collection process.

Equine activity (sector)	Norway	Sweden
Riding schools	Survey (71 respondents)	Survey (99 respondents)
Trainers	Survey (54 respondents)	Interviews
Breeders	Survey (171 respondents)	Survey (104 respondents)
Boarding ###enterprises?	Survey (190 respondents)	Gross margin budgets[1]
Tourism	Survey (25 respondents)	Not pursued

[1] Information on gross margin budgets was obtained from Agriwise (http://www.agriwise.org/), which is a planning tool at the Swedish University of Agricultural Sciences.

The FLQ formula takes the following form:

$$FLQ_{ij} = CILQ_{ij} \times \lambda^*$$

$$\lambda^* = [\log_2 (1 + TRE/TNE)]^\delta \text{ and } 0 \leq \delta < 1$$

where TRE and TNE are total regional and national employment, respectively, and $\log_2(X)$ is equivalent to $\ln X/\ln 2$ and represents a concave function between [0-1]. The parameter δ will determine the actual shape of the relationship between λ^* and TRE/TNE and for values of $\delta<1$, the function λ^* will increase at a decreasing rate as TRE/TNE approaches unity. In numerical tests based on data from the UK (Flegg and Webber, 2000) and Finland (Flegg and Tohmo, 2010), it has been shown that a value of $\delta=0.3$ gives the best regional results in comparison with survey-based estimates. It should be emphasised that the FLQ correction for regional size is general for all cells of the IO table and that it does not capture regional specialisation more than already done by the CILQ. The purpose of the correction is rather to scale down technical coefficients, as it is postulated that a negative relationship exists between regional size and self-sufficiency.

Results

The highest multipliers in Sweden were found for riding schools and breeders (Table 3). The reason for such high multipliers may be that these enterprises (often in the form of a voluntary and hobby activity) are not making any profits and all revenue is used for purchasing inputs to the activities and hence spent in the supply chains. Activities needing to allocate some revenue to return to capital and wages showed somewhat lower multipliers, e.g. boardingl enterprises and professional trainers.

For the national Swedish IO model, the multipliers for the horse sectors ranged between 1.47 and 2.15 for the open model and 2.61 and 3.19 for the closed model (Table 3). This can be interpreted such that a unit (1) increase in demand for such products has the corresponding multiplier impact on the entire national economy. This is the effect on the sector itself, and on all other sectors in the economy due to the scaling up of production in step-after-step. When we compared these to some other multipliers for the agricultural sector, we found that they were in line with these values, and that both breeders and riding schools display multipliers in the same high magnitude as e.g. intensive animal farms. It is interesting to note that when we compared these results to the sector of sporting and recreation services, the horse sectors displayed higher multipliers than this category of activities. We also report results for some other sectors of the economy in Table 3 to allow for comparisons.

Turning to the regional results (Table 4), it is evident that multipliers are high in such regions where the surrounding economy can support horse-related enterprises and activities and provide feed, equipment, transport services, etc. Even if the number of horses per capita is impressive on the island of Gotland, the multiplier impact of horses on the surrounding economy was rather weak (Table 4). The interpretation is that as the region is rather small, and also characterised by being an island, there is a large amount of imports from other Swedish regions (or from abroad) in inputs to the horse sector. However, this is also true for e.g. agricultural activities, which show low regional multipliers. In comparison, the large and dense regions of Stockholm, Skåne and Västra Götaland displayed higher multipliers, indicating that horse activities are integrated into the local economies (as are many sectors of such regional economies). Värmland is included in Table 4 to show the magnitude of multipliers in a somewhat remoter region, but with larger urban centres and more regional production of feed and horse-related products. This region is similar to the Buskerud region in Norway (Table 6).

The results for Norway were rather similar to those for Sweden, with the exception that the multiplier for trainers was higher in Norway (Table 5). For Norway, we were also able to model the horse

Table 3. Results from the Swedish national input-output analysis.

Sector	National Swedish multipliers		
	Open model	Closed model	Emp./mill. SEK[1] closed model
Trainers	1.74	2.61	2.02
Boarding enterprises	1.47	2.86	2.12
Breeders	2.00	2.90	2.74
Riding schools	2.15	3.19	2.91
Milk	1.92	2.78	2.72
Cattle/deer	1.91	2.85	3.03
Pigs	1.95	2.57	1.80
Poultry/eggs	2.24	2.92	1.93
Sheep	1.86	3.11	NA[2]
Cereals	1.76	2.77	3.30
Forage	1.80	2.28	1.31
Other crops (potato/sugar)	1.81	2.44	1.89
Agricultural services	1.80	2.89	3.62
Food products, beverages and tobacco products	1.90	2.52	1.46
Textiles, wearing apparel and leather products	1.68	2.39	1.63
Paper and paper products	1.87	2.42	1.10
Electrical equipment	1.70	2.33	1.23
Constructions and construction works	1.72	2.57	1.78
Sporting services and amusement and recreation services	1.70	2.52	2.36

[1] Emp./mill. SEK = employment numbers/million Swedish Kronor (SEK). In 2012 1 SEK = 0.1149 Euros
[2] NA: not available.

tourism sector, which had open multipliers of a magnitude in line with e.g. riding schools and boarding enterprises, but closed multipliers with lower values than riding schools (indicating less labour-intensive activities) but higher than lboarding (being more intensive than the latter). The horse-related multipliers in Norway were also rather in line with other recreational activities, both for open and closed multipliers. However, the employment impact seemed to be higher for horse-related activities compared with sporting and recreational activities.

For the Norwegian region of Buskerud (Table 6)[5],we found that the multipliers were similar to those for Värmland. Riding schools and trainers displayed high multipliers, whereas boarding seemed to be an activity with sparse use of inputs from other economic sectors. The employment impacts of increases in horse activities seemed to be similar to those of agriculture, although somewhat lower. Hence a shift from agriculture to horses in the region seemed to be rather neutral for the economic upstream impacts on the regional economy, but the employment impacts would be marginally to moderately negative depending on the type of activity. Boarding creates the fewest jobs, whereas riding schools create the most.

---

[5] Buskerud is the only region in Norway for which we were able to develop a IO model at the regional level.

Table 4. Results from the Swedish regional input-output analysis.

Sector/Region   Regional Swedish multipliers

Sector/Region	Gotland[1]		Skåne[2]		Stockholm[3]		Värmland[4]		Västra Götaland[5]	
	Open	Closed	Open	Closed	Open	Closed	Open	Closed	Open	Closed
Trainers	1.03	1.03	1.28	1.33	1.37	1.48	1.14	1.15	1.37	1.44
Boarding enterprises	1.02	1.02	1.16	1.25	1.22	1.43	1.10	1.11	1.23	1.36
Breeders	1.04	1.04	1.36	1.40	1.50	1.62	1.19	1.20	1.48	1.55
Riding schools	1.06	1.06	1.40	1.45	1.54	1.67	1.23	1.24	1.56	1.64
Milk	1.03	1.03	1.38	1.42	1.50	1.61	1.20	1.20	1.42	1.48
Cattle/deer	1.04	1.04	1.35	1.40	1.49	1.61	1.12	1.13	1.42	1.49
Pigs	1.03	1.04	1.20	1.23	1.48	1.55	1.18	1.19	1.44	1.48
Poultry/Eggs	1.08	1.08	1.32	1.34	1.47	1.54	1.23	1.23	1.55	1.59
Sheep	1.02	1.02	1.34	1.41	1.42	1.59	1.13	1.14	1.40	1.50
Cereals	1.03	1.03	1.17	1.22	1.42	1.56	1.13	1.14	1.37	1.45
Forage	1.06	1.06	1.32	1.34	1.43	1.48	1.06	1.07	1.38	1.41
Other crops	1.04	1.04	1.13	1.15	1.42	1.50	1.19	1.19	1.40	1.45
Agricultural services	1.05	1.05	1.25	1.32	1.39	1.54	1.15	1.16	1.38	1.46

[1] Large (rural) island on the east coast of Sweden.

[2] Important agricultural region in southern Sweden with many large cities.

[3] Capital region.

[4] Rural region in mid-Sweden on the border to Norway, with some medium size cities.

[5] Large region in south-west Sweden with many urban areas and the second largest city (Gothenburg).

Discussion and concluding remarks

Equine activities have increasingly gained in importance in affluent economies over the last twenty years and, most likely reflecting this trend, policy makers have shown growing interest in acknowledging the importance and role of this sector. This has led relevant academic institutions and consultancy firms to meet this interest by providing economic impact analysis of horse-related activities at the national, regional and local level. In most cases, the analytical tool used in these studies is based on an IO model where multiplier effects are first determined and then used to infer the economic importance of equine activities in terms of output, value added and employment for the national, regional or local economy concerned.

In this analysis of the economic impacts of horse-related activities in Norway and Sweden, we broke the figures down to the regional level to show the difference between national and regional impact coefficients if the sector were to expand. We applied the IO model to examine such linkages and impacts. At present, horse-related information is either lacking or distributed across different sectors of the economy in national IO accounts in both Sweden and Norway. To deal with this issue, we examined earlier attempts to create a separate horse sector in IO models and devised a simple method for disaggregating such accounts based on different sources of data, including surveys, interviews and disaggregated sector data. As shown in the results section, we developed IO coefficients for

Table 5. Results from the Norwegian national input-output analysis.

Sector	National Norwegian multipliers		
	Open model	Closed model	Emp./mil. NOK[1] closed model
Riding schools	1.76	3.51	2.74
Breeders	1.91	2.96	2.58
Horse tourism	1.74	2.65	2.32
Trainers	2.04	3.22	2.27
Boarding enterprises	1.46	2.28	1.64
Agriculture, hunting and related services	1.97	2.61	2.38
Food products, beverages and tobacco products	2.32	3.07	1.54
Textiles, wearing apparel and leather products	1.77	2.61	1.60
Paper and paper products	2.05	2.87	1.36
Constructions and construction works	1.88	2.78	1.53
Accommodation and food services	1.72	2.67	1.91
Sporting services and amusement and recreation services	1.69	2.59	1.56

[1] Emp./mil. NOK = employment numbers/million NOK. 1 Norwegian kronor (NOK) = 0.1339 Euros.

Table 6. Results from the regional Norwegian input-output analysis.

Sector	Regional Buskerud[1] multipliers		
	Open model	Closed model	Emp./mil. NOK[2] closed model
Riding schools	1.22	1.25	1.59
Breeders	1.10	1.11	1.50
Horse tourism	1.10	1.11	1.49
Trainers	1.19	1.21	1.15
Boarding enterprises	1.06	1.07	1.03
Agriculture, hunting and related services	1.18	1.19	1.63
Forestry, logging and related services	1.10	1.11	1.12
Food products, beverages and tobacco products	1.23	1.24	0.53

[1] Region directly west of Oslo with both urban centres and remote rural mountainous areas.

[2] Emp./mil. NOK = employment numbers/million NOK. 1 Norwegian kronor (NOK) = 0.1339 Euros.

horse-related activities in both countries and used these to derive multipliers showing the impacts on the economy of expanding different horse activities. We also examined the regional structure of the multipliers.

Regarding Sweden, the results indicate that the highest multipliers exist for riding schools and breeders. The reason for such high multipliers can be that these enterprises use most of the revenue for purchasing inputs and spend it in the supply chains, and hence do not make much profit. Activities which typically need to allocate some revenues to returns on capital and wages have somewhat lower multipliers, e.g. boarding enterprises and professional trainers. At the regional level, open and closed output multipliers in Sweden are smaller than their national counterparts. This can be expected due to the greater openness of regional economies than the corresponding national economy. Another relevant finding at the regional level in Sweden is that more populated and/or urban regions such as Stockholm and Skåne display multipliers for horse-related activities that are greater than those obtained for other Swedish regions, which tend to be more rural. As far as Norway is concerned, the multiplier results obtained are comparable to those obtained for Sweden. In addition, it proved possible to obtain output multipliers for horse tourism in Norway which were similar in magnitude to those generated for recreational activities such as riding schools.

A multiplier analysis of this kind does not provide any particular evidence that focusing on horse-related activities would stimulate the regional economy more than e.g. agriculture. Hence, it should not be viewed as a development strategy purely based on economic stimulation of rural areas. However, there might be other aspects such as the interlinkages between urban and rural areas, cultural values or landscape values which motivate horse activities as a strategy for regions. Moreover, as more and more areas are becoming dominated by horse activities as agriculture contracts, it is interesting to know that the economic impacts for the regions are similar from expansion of horse activities as they would be from agriculture. Obviously this analysis was based on the backward supply chains, but the impact on forward linkages could also be calculated and analysed.

The empirical work reported in this study can be extended in several directions and can be viewed as preliminary in a European context. Thus, social accounting matrices (SAM) constructed at the local or regional levels can be a potential avenue to analyse the spatial interactions between rural and urban areas associated with horse-related activities (Psaltoppoulos *et al.*, 2006; Roberts, 2000). Furthermore, this application of IO analysis to Scandinavian countries could be of interest to analysts in other European countries seeking to assess the role of horse-related activities. It could be especially relevant for some regions of France and the United Kingdom, where the horse sector is an important contributor to regional and local economies.

Acknowledgements

The authors gratefully acknowledge financial support received from the Swedish-Norwegian Foundation for Equine Research under the project H0947227 entitled 'Economic growth potential in the Norwegian and Swedish equine sectors in a national and regional perspective'.

References

Andersson, H. and D. Johansson, 2004. The horse sector: does it matter for agriculture? Annual Meeting of the American Agricultural Economics Association, Denver, CO, USA, 21 pp.

ATG, 2011. Swedish Horse Racing Totalisator Board. Annual Report, Sweden.

Beattie, B.R., T. Teegerstrom, J. Mortensen and E. Monke, 2001. A partial economic impact analysis of Arizona's horse industry. Project completion report, phase II: horse show impact. Department of Agricultural and Resource Economics, University of Arizona, Tucson, AZ, USA, 43 pp.

British Horseracing Authority, 2013. Economic impact of the British racing. London, United Kingdom, 60 pp. Available at: http://tinyurl.com/p58gcng.

Canmac Economic Limited, 2008. Nova Scotia harness racing industry economic impact study. Lower Sackville, Canada, 52 pp. Available at: http://tinyurl.com/onxrjjp.

Econ Pöyry, 2008. Samfunnsregnskap for travhest og galopp *(Socioeconomic analysis for trotter and gallop)*. Oslo, Norway.

Econometric Research Limited, 2001. The economic impacts of horse racing and breeding in Alberta. Alberta Horse Racing Industry Review Working Committee, Edmonton, Canada.

Econometric Research Limited, 2011. The economic impacts of horse racing and breeding in Ontario 2010. Review for the Ontario Horse Racing Industry Association, Toronto, Canada.

Flegg, A.T., C.D. Webber and M.V. Elliott, 1997. On the appropriate use of location quotients in generating regional input-output tables: reply. Regional Studies 31(8): 795-805.

Flegg, A.T. and C.D. Webber, 2000. Regional size, regional specialization and the FLQ formula. Regional Studies, 34(6): 563-569.

Flegg, A.T. and T. Tohmo, 2010. Regional input-output tables and the FLQ formula: a case-study of Finland. Regional Studies, 47(5): 703-721.

Gren, I., 2006. Översyn över hästspelsmarknaden i Europa (*Overview of the market for horse gambling in Europe*). ATG (Swedish Horse Racing Totalisator Board) and HNS (The Swedish Horse Council Foundation), Sweden.

Hughes, D., 2003. Policy uses of economic multiplier and impact analysis. Choices, second quarter, 25-29.

Hughes, D.W., J.A. Woloshuk, A.C. Hanham, D.J. Workman, D.W. Snively, P.E. Lewis and T. Walker, 2011. An evaluation of the impact of the equine industry on the West Virginia economy. Research report, West Virginia University Extension Service, West Virginia University, Blackburg, VA, USA, 66 pp.

Leontief, W., 1966. Input-output economics. Oxford University Press, New York, NY, USA.

Liljenstolpe, C., 2009. Horses in Europe. Equus. Swedish University of Agricultural Sciences, Report, 26 pp.

Lindberg, G., 2011. Linkages: economic analysis of agriculture in the wider economy – input – output models and qualitative evaluation of the common agricultural policy. SLU services, Upssala, Sweden.

Lindall, S.A. and D.C. Olson, 2008. The IMPLAN input-output system. IMPLAN 2.0 IO System Description. Minnesota Implan Group Inc. Stillwater, MN, USA.

Menard, R.J., K.W. Hanks, B.C. English and K.L. Jensen, 2010. Tennessee's equine industry: overview and estimated economic impacts. Staff paper 10-01. Department of Agricultural Economics, University of Knoxville, Knoxville, TN, USA, 35 pp.

Midmore P. (ed) 1991. Input-output models in the agricultural sector. Avebury Academic Publishing Group, Brookfield, VT, USA.

Midmore, P. and L. Harrison-Mayfield (eds) 1996. Rural economic modelling, an input-output approach. CAB International, Wallinford, United Kingdom.

Miller, R.E. and P.D. Blair, 2009. Input-output analysis: foundations and extensions. Cambridge University Press, Cambridge, United Kingdom.

Organisation for Economic Co-operation and Development (OECD), 2009. The role of agriculture and farm household diversification in the rural economy: evidence and initial policy implications. OECD, Paris, France.

Psaltopoulos, D., E. Belamou and K.J. Thomson, 2006. Rural-urban aspects of the CAP measures in Greece: An inter-regional SAM approach. Journal of Agricultural Economics 57(3): 441-458.

Roberts, D., 2000. Spatial diffusion of secondary impacts: rural-urban spillovers in Grampian Scotland. Land Economics 76(3): 395-412.

Ten Raa, T., 2005. The economics of input-output analysis. Cambridge University Press, Cambridge, United Kingdom.

University of Kentucky, 2013. 2012 Kentucky equine survey final report and technical appendices. College of Agriculture, Food and Environment, University of Kentucky, Lexington, KY, USA, 19 pp. Available at: http://equine.ca.uky.edu/kyequinesurvey.

3. The 2012 Kentucky Equine Survey: importance and impact of the equine industry in Kentucky

R.J. Coleman[1], M.G. Rossano[1], C.J. Stowe[2]*, S. Johnson[2], A.F. Davis[2], J.E. Allen IV[2], A.E. Jarrett[2], G. Grulke[3], L. Brown[4] and S. Clark[4]

[1]Department of Animal and Food Sciences, University of Kentucky, 613 W.P. Garrigus Building, Lexington, KY 40546-0215, USA
[2]Department of Agricultural Economics, University of Kentucky, 307 Charles E. Barnhart Bldg., Lexington, KY 40546-0276, USA; jill.stowe@uky.edu
[3]Kentucky Horse Council, 1500 Bull Lea Rd, Lexington, KY 40511, USA
[4]USDA-NASS, P.O. Box 1120, Louisville, KY 40201, USA

Abstract

Little is formally known about the U.S. equine industry. Little is also known about Kentucky's equine industry, even though state bills itself as the 'Horse Capital of the World,' with the last comprehensive study being conducted in 1977. However, having reliable data is crucial to the success and sustainability of the industry. The 2012 Kentucky Equine Survey was conducted to provide data for decision-making. The objectives of the study were to obtain an inventory of all equine (Phase 1) and estimate the industry's economic impact (Phase 2). Phase 1 of the study was conducted in conjunction with the National Agricultural Statistics Service (NASS). Phase 2 of the study utilized data collected by NASS as well as data from a set of supplementary surveys. Results from this study will be used to identify areas for growth within the industry and opportunities for rural development.

Keywords: Kentucky equine industry, equine operation inventory, economic impact analysis

Introduction

As with any key industry, the equine industry prospers when it is measured and assessed. To this end, it is critical to have a current inventory of the existing population of horses and equine operations as well as to understand the economic impact of the industry. And in the United States, while Kentucky promotes itself as the 'Horse Capital of the World,' the last thorough study of Kentucky's equine industry was undertaken in 1977 (1977 Kentucky Equine Survey, 1978), leaving the industry with 35 years of changes that were neither measured nor analyzed. In addition, with ever-increasing market volatility, having reliable data upon which to base decisions is crucial to the future success and sustainability of this industry.

Kentucky's equine industry extends far beyond just horses. The equine industry offers a number of unique, equine-related tourism opportunities. There are ancillary supporting businesses such as veterinarians and farriers, as well as providers of fencing, feed, bedding, insurance, laundry services, loans, pharmaceuticals and specialized educational opportunities, among others. Some have even characterized Kentucky's equine industry, particularly in the Central Bluegrass region, as an economic cluster, along with more familiar economic clusters like Napa Valley (wine) and Silicon Valley (tech/computers) (Garkovich et al., 2008; Porter, 2000).

In response to these needs, the University of Kentucky (UK) and the Kentucky Horse Council (KHC), with support from numerous industry organizations, succeeded in securing funding in 2011 for a large-scale, comprehensive equine survey upon which future industry plans could be built.

Materials and methods

The Kentucky Equine Survey consisted of two phases. Phase 1 of the study was a statewide survey of equine operations that included an inventory of all breeds of equine, including horses, ponies, donkeys, and mules, as well as sales, income, expenses and assets of those operations. An equine operation was defined as an address where at least one horse, pony, donkey, or mule resided. All results were reported at the county level. Phase 2 of the project was an economic impact analysis of Kentucky's equine industry.

Phase 1: equine operations inventory

Phase 1 was conducted between July and October 2012. The inventory of equine operations was conducted by the Kentucky Field Office (now the Eastern Mountain Region Office) of the National Agricultural Statistics Service (NASS), which is an agency of the United States Department of Agriculture (USDA). NASS followed a list-segment and area-segment procedure.

When the study began, no current or comprehensive list of equine operations that included operations not fitting the NASS definition of a farm existed for Kentucky (according to NASS-USDA, a farm is any establishment that has at least $1000 in cash receipts annually). In addition, the long period of time since the previous equine survey necessitated a communications campaign to inform members of the equine industry about the survey, why it was beneficial, and ensuring individuals that responses were confidential. List building for the survey sample involved acquiring names and addresses of members of cooperating equine organizations and a general solicitation for individuals to submit contact information through a web page hosted by KHC. Additional names and addresses were collected at a series of 36 public engagement meetings. These meetings were programmed by UK faculty and extension workers in counties around the state. In addition to a presentation about the survey, an educational program about a horse-related topic was usually provided to motivate audience attendance. During meetings, attendees were encouraged to provide contact information to the survey personnel.

The list building efforts resulted in the collection of 13,059 names and addresses. Names were checked for duplication against those already on the NASS list and duplicates were removed. NASS contacted a portion of operations by telephone to obtain preliminary information regarding the numbers of horses at those operations so that the survey sample could later be stratified by size according to the approximate number of horses. The final list was comprised of operations and individuals including private owners of one to two horses at their residences, boarding facilities, large commercial breeding operations, and race tracks. From the entire list, a random sample, stratified by geographic location and size, was drawn, and surveys were sent to 15,000 equine operations. If surveys were not returned, telephone enumerators contacted the operations to obtain the information. In addition, field enumerators visited some of the largest farms included in the study to assist with data collection. To capture information on equine operations not on the final list, the equine survey was included in the Agricultural Coverage Evaluation Survey, which was combined with the June Area Survey sample in constructing the area component of the sample. Two hundred seventy-nine segments of land were canvassed by field enumerators who collected data on all agricultural activities in those areas.

Of the 15,000 surveys distributed, 10,753 (72%) produced responses. Of those, 1,042 refused to participate; the remaining 9,711 records were used for analysis. Surveys from operations with at least one equid were reviewed, edited and entered into a database by NASS personnel. When a survey was partially completed or the non-respondent was an extremely large operation, imputation was utilized to account for non-response. Otherwise, non-response was accounted for through an adjustment to the original sampling weights. List sample records were expanded by strata and summarized;

then, records from the 279 area segments that were not on the list (NOL) were expanded and added to the results of the list to produce state-level multiple frame indicators. To produce more robust county-level indicators, a final reweighting was then done, by which weights on NOL records were set to zero while weights on list records were adjusted, such that the expanded state list indication equaled the expanded state multiple frame indication. The list sample records were expanded by this final weight to produce county-level indications. The estimation process produced an estimate of total equine in the state with a relative error of approximately 1.2% of the estimate. This figure was determined by the size and variance of the sample, the proportion of the population sampled, the response rate, and the sampling weights.

Phase 2: economic impact analysis

Phase 2 was conducted between June 2012 and April 2013. A variety of economic measures of Kentucky's equine industry was estimated through the use of data from the NASS inventory study using IMPLAN as well as data collected in a set of supplementary surveys. The methodology for each of these is described below.

IMPLAN analysis

To estimate the economic impact of the equine industry on Kentucky, income and expenditure data from the NASS inventory study, as well as supplementary data from studies described below, were utilized in an input-output (IO) model with 2011 IMPLAN data. Briefly, IO modeling is an approach that considers the interdependencies between economic sectors and estimates the multiplier effect from additional business and household spending within a predefined geography (Miller and Blair, 2009). IMPLAN (IMpact analysis for PLANning), originally developed in 1972 by the U.S. Forest Service, is the most common methodology to assess economic impact and is widely used across industries (www.implan.com).

Economic impact was measured in three ways: (1) the output effect, which measures the increase in sales due to the presence of an industry; (2) the employment effect, which measures the number of jobs generated as a result of the presence of the equine industry; and (3) the value added effect, which measures new income paid to workers, profits earned by businesses, or dividends paid to shareholders.

In each measure, the full economic impact of the equine industry includes the 'multiplier effect,' which summarizes the total impact that can be expected from a change in a given economic activity. For example, a new breeding facility represents an economic change which can spur and support new spinoff activities, such as veterinary or transportation services. Multipliers measure the economic impact of these new products or services, including the resulting spinoff activities.

While there are several types of multipliers, the Type II multiplier is most widely used in IO analysis. A Type II multiplier includes the effect of *direct* or initial spending, *indirect* spending or businesses buying and selling to each other, and household spending based on the income earned from the direct and indirect effects. Essentially, these latter *induced* effects represent employee spending on goods and services. The multiplier is calculated by dividing the sum of direct, indirect, and induced effects by direct effect; therefore, if there are any indirect or induced effects, the multiplier will be greater than one.

Separate multipliers were estimated for each of the equine sectors and Type II spending types; the complete set of multipliers is available from the corresponding author upon request. The average equine industry multipliers, weighted based on the revenue contribution of each equine sector to the whole industry, is 1.68 (output), 1.74 (value added), and 1.40 (employment). The latter statistic suggests, for example, that for every 10 jobs created in the equine industry, four jobs are created outside of the industry in spinoff activities. The ranges of the sector multipliers used in the calculations are 1.4-2.0 (output), 1.3-2.9 (value added), and 1.2-5.2 (employment).

Event attendance surveys

Equine-related events generate a number of tourism impacts. Competitors, family, friends, and pure spectators attend these events, spending money on the event grounds, in the local community, and in the state. Total attendance at equine events was not tracked, but per-person spending at events was estimated. A team of researchers from the Kentucky Equine Survey attended 11 events and four race meets, intercepting attendees throughout the day of the event. Attendees were asked to complete a questionnaire which included demographic information as well as questions related to spending in different categories in different areas of the state.

All attempts were made to broadly sample different types of equine events, including major horse shows, smaller horse shows, and organized trail rides. In addition, the study team visited race meets across the state. A total of 1,722 surveys were collected from 15 different events held between June and November 2012.

Using the data from these surveys, estimates of per-person spending at each event as well as total economic impact from the event were calculated, categorized by event size. Even though the primary purpose of the visit was taken into account, only out-of-state visitors were included in the analysis because they represent new money entering the Kentucky economy.

Racetrack management survey

A survey was sent to the general manager or owner of six racetracks in Kentucky: Churchill Downs, Ellis Park, Keeneland, Kentucky Downs, Red Mile, and Turfway Park. The survey requested information from the calendar year 2011 in the following categories: revenues, operating expenses, assets, capital investments, investments in human capital and technology, and use of wagering technology. Only one survey (16.7%) was returned. Consequently, all needed racetrack data was obtained from the Kentucky Horse Racing Commission's 2010-2011 biennial report.[1]

Non-market valuation survey

The non-market valuation study was conducted to obtain an estimate of the value of the externalities generated by the presence of the equine industry in the state of Kentucky, which may include recreational, environmental, and aesthetic benefits. The writing and administration of the survey was accomplished in four stages. First, a preliminary draft of the survey was created to closely replicate the survey used in Ready (1991). Second, a focus group was conducted to examine the effectiveness, clarity, and navigability of the survey instrument. Third, the final draft of the survey was prepared; 6,176 surveys were compiled and mailed to residents of eight counties (Bourbon, Clark, Fayette, Harrison, Jessamine, Madison, Scott, and Woodford) in the Bluegrass Region of Kentucky. An additional 2,000 surveys were mailed to randomly selected Kentucky residents outside of the Bluegrass Region.

To recruit the survey sample, Fayette County residents were randomly selected from a database obtained from the Fayette County Property Valuation Administrator. Addresses for residents outside of Fayette County were obtained from the marketing company USA Data. Weeks after the survey mailing, reminder postcards were mailed to increase response rate. The overall response rate, after

---

[1] This report can be accessed at: http://tinyurl.com/nh9jd9f.

accounting for bad addresses, was 22.28%. Response in the Bluegrass Counties was significantly higher than non-Bluegrass Counties (25.19 vs 10.10%).

Results

Equine operations inventory

Estimates from this study suggest that there are a total of 35,000 equine operations throughout Kentucky. These equine operations accounted for a total of 4.3 million acres of land, of which 1.07 million acres were devoted to equine-related activities.

Farms or ranches comprised the largest component of equine operations by primary business type. The survey results showed that 19,500 Kentucky equine operations (56%) were listed as farms or ranches, and 10,500 operations (30%) were properties on which equine were kept for personal use. In addition, 1,100 operations (3%) were listed as a boarding, training or riding facility, and 700 (2%) were listed as a breeding operation. Finally, another 3,200 (9%) operations were identified as 'other,' which could include facilities like therapeutic riding centers, equine rescue operations or some type of non-equine operation not identified above.

Survey respondents were asked to estimate the percentage of equine-related activities on their operation that were conducted for business purposes (i.e. to generate income). These responses differed significantly according to primary type of the operation. Breeding operations had the highest percentage of activities that are income-generating (73%), followed by boarding, training or riding facilities (56%). Farms or ranches indicated that 10% of activities were for business purposes, followed by 'other' (8%) and operations on which equine are kept for personal use (6%). So, about 95% of the equine operations (farms and ranches, personal use, and other) have 10% or less of their equine-related activities for business purposes, indicating a significant recreational segment.

There were an estimated 242,400 horses, ponies, mules and donkeys in the state of Kentucky on July 1, 2012. The majority of the state's equine are light horse breeds, which accounted for 89.2% of the total. Donkeys and mules accounted for 5.8% of the state's equine, with ponies accounting for 2.9% and draft horse breeds accounting for 2.1%. Figure 1 illustrates the distribution of horses

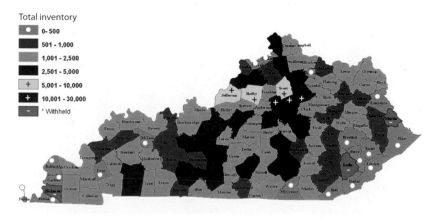

* Withheld to avoid disclosing data for indiviual firms

Figure 1. Kentucky equine inventory by county as of July 1, 2012.

around the state. The most densely populated counties are Bourbon, Fayette, and Woodford, followed by the corridor from west to east including Jefferson, Shelby and Scott counties.

Figure 2 and 3 illustrate inventory by breed and primary use. The three most prominent breeds are Thoroughbreds (54,000), Quarter Horses (42,000), and Walking Horses (36,000); the top three

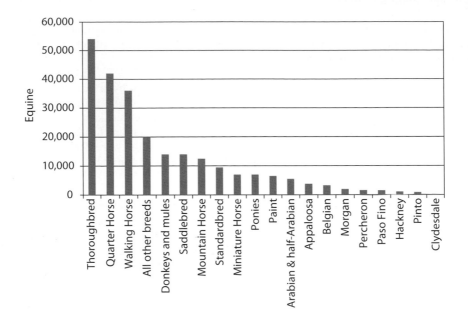

Figure 2. Inventory of equine by breed as of July 1, 2012.

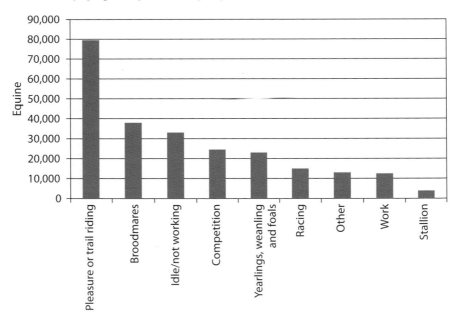

Figure 3. Inventory of equine by primary use as of July 1, 2012.

primary uses of equine in Kentucky are pleasure or trail riding (79,000), broodmares (38,000), and idle or not working (33,000).

The value of all horses, ponies, mules and donkeys was $6.3 billion on July 1, 2012. The estimated total value for Thoroughbred horses was the highest ($5.5 billion), followed by American Saddlebreds ($173 million), Quarter Horses ($146 million), Standardbreds ($119 million), Walking Horses ($71 million) and Mountain Horse breeds ($31 million).

Many assets, capital and otherwise, are required to support equine and equine-related activities. The reported value of equine-related land, fencing and buildings was $13.9 billion. The value of equine-related vehicles and equipment was reported to be almost $3 billion. The reported value of tack and equestrian clothing was $114 million, and the value of feed and supplies was $58 million. In total, the value of all equine-related assets, including the value of the equine in the state, was estimated at $23.4 billion.

Income from sales of horses, ponies, mules and donkeys was $522.1 million. Income from equine-related services was $491 million, with $220 million coming from breeding services and $271 million coming from non-breeding services. These figures suggest a dynamic industry, encompassing a broad range of activities, including breeding, preparing horses for the market place, recreational use and equine and human health endeavors. In total, equine-related income from sales and services for equine operations in 2011 was estimated to be about $1.1 billion.

Total equine-related expenditures in Kentucky in 2011 amounted to $1.2 billion. Capital expenditures, which at $338 million accounted for 29% of all equine-related expenditures, include the purchase of real estate, horses, ponies, mules and donkeys, and expenditures on improvements and equipment. Operating expenditures, excluding labor, include many items, such as board, feed, bedding, veterinarian and farrier services, supplies, tack and equipment, breeding fees, maintenance and repair, insurance premiums, utilities and fuel, taxes, rent or lease, fees and payments, transportation, training fees and other expenses. In 2011, operating expenditures, excluding labor, totaled $839 million. Notably, 77% of these operating expenses were spent in Kentucky.

There were a total of 13,000 workers reported on Kentucky's equine operations during 2011, of which 5,600 were full-time and 7,400 were part-time. The total payroll expenses paid to these workers was estimated at $202 million. Additionally, non-wage benefits, which include housing, utilities, meals, clothing, transportation, horse boarding and riding lessons, totaled an estimated $16.7 million during the same period.

Economic impact analysis

The economic impact of the equine industry was measured in three ways: the output effect, the employment effect and the value added effect. Direct economic impacts are those actually generated within the industry, while the total economic impact multiplies the direct impact by the industry multipliers to estimate the impact of direct, indirect (i.e. spinoff) and induced activities.

Total economic impact estimates are presented in Table 1. When considering only direct effects, the entire equine industry is responsible for generating over 32,000 jobs (which includes the 13,000 jobs directly on equine operations plus jobs in other ancillary businesses, such as feed producers, breed and discipline organizations, tack stores, and so on) and $1.78 billion in new revenue. When accounting for direct and spinoff activities, the equine industry in Kentucky is responsible for over 40,000 jobs and nearly $3 billion in new revenue.

Table 2 presents these employment, output, and value added estimates by industry sector. According to the results, breeding is the most labor-intensive sector, and the racing sector contributes the most to new revenue in the state.

State and local tax revenues are derived from all of the direct and indirect activities generated by the equine industry. In addition, the equine industry contributes an above average share of sales tax to Kentucky compared to the rest of the agricultural industry because, unlike the rest of the state's animal agriculture (i.e. beef cattle, dairy cattle, swine, poultry and goats), equine farm purchases are not exempt from Kentucky sales tax. It is estimated that the Kentucky tax impact of the equine industry is approximately $134 million annually, which includes tax revenue collected from state income tax and sales tax, but does not include occupational license tax revenues.

Kentucky's equine industry is a rich source of tourism revenue. For example, horse shows generate additional sales for the hospitality and transportation industries through hotel rooms, eating establishments, gasoline purchases and other sundry expenditures at drugstores, apparel stores and entertainment venues such as movies and vineyards. Table 3 provides a guideline for economic impacts at horse shows of different sizes. A guideline for racetracks was not included because the data collected were not sufficiently representative to compute a legitimate estimate of economic impact.

Non-market valuation

The value of Kentucky's equine industry extends beyond tourism impacts and transactions in the marketplace. There are many benefits to a community which is home to a healthy horse economy.

Table 1. Economic impact of Kentucky's equine industry.

	Employment (jobs)	Output (million US$)	Value added (million US$)
Direct equine industry impacts	32,022	1,780	813
Total equine industry impacts	40,665	2,990	1,400

Table 2. Estimate of impact by sector.

Sector	Employment (jobs)	Output (million US$)	Value added (million US$)
Breeding	16,198	710	333
Competition	2,708	635	297
Racing	6,251	1,280	601
Recreation	594	166	78
Other	14,914	194	91

Table 3. Economic impact of horse shows in Kentucky.

Size of show	# of attendees	# of out-of-state attendees	Total economic impact (US$)
Small	200	80	21,400
Medium	500	200	53,600
Large	2,000	800	214,400

Some benefits cannot be directly measured, but are just as real as the economic impacts. There is value to both area and non-area residents in the existence of the rural landscapes created by equine operations, which may be recreational, environmental and/or aesthetic, and as such, there is a value to preserving these types of landscapes created by the presence of the horse industry.

A large majority of respondents (87.3%) were supportive of a free, hypothetical program which would protect agricultural farm land, including equine operations, from development. Of those that were supportive, 82.2% were willing to pay some amount of increased taxes every year to preserve the equine industry at its current size; notably, 59.6% were willing to pay at least $50 a year.

Conclusion

The 1977 Kentucky Equine Survey estimated that there were 204,000 equine in Kentucky, compared to the 242,400 equine from the 2012 Kentucky Equine Survey. The 1977 study did not estimate economic impact. Because of 35-year time span in between, it would be futile to hypothesize on the evolution of equine population numbers over time.

In addition to better understanding the current state of the equine industry in Kentucky, it is also beneficial to compare the economic impact of the equine industry to other agricultural commodities in the state that were studied recently using similar methodology. In 2012, the corn growing industry in Kentucky showed an output effect of $1.27 and an employment effect of 27,313 jobs (Kentucky Agriculture Facts, 2013). A study for the Kentucky Soybean Association showed an output effect of $1.36 billion, a value added effect of $649 million, and an employment effect of 21,980 jobs (A.F. Davis, personal communication, 2013). Finally, a study of the forest and wood industries in Kentucky, including associated forestland, estimated an output effect of $9.9 billion and an employment effect of 51,928 jobs; no value added effect was estimated (Stringer *et al.*, 2013). Although no recent studies have been conducted on the state's beef, dairy, poultry, coal, or bourbon industries, which are all significant commodities in the state, the above comparisons show that the equine industry is a major contributor to the state's agricultural economy.

The hope is the 2012 Kentucky Equine Survey can serve as a benchmark for future studies and applied work relating to the equine industry. The methodology used is applicable to all other states and countries which have access to an association like NASS; a description of the methodology was provided in Stowe and Rossano (2013). Furthermore, with these results, Kentucky policy makers have better data on which to make decisions, entrepreneurs and business owners have better data from which to develop business plans and ideas, veterinarians have the ability to better monitor disease outbreaks, community planners can better facilitate future projects, and the state has a baseline going forward as the industry navigates the 21st century.

References

Garkovich, L., K. Brown and J.N. Zimmerman, 2008. 'We're not horsing around': conceptualizing the Kentucky horse industry as an economic cluster. Community Development Journal 39(3): 93-113.

Kentucky Agriculture Facts, 2013. Kentucky Farm Bureau, Louisville, KY, USA, 27 pp.

Kentucky Horse Council, 1978. Kentucky equine survey 1977. Kentucky Horse Council, Lexington, KY, USA, 52 pp.

Miller, R.E. and P.D. Blair, 2009. Input-output analysis: foundations and extensions. Cambridge University Press, New York, NY, USA, 784 pp.

Porter, M.E., 2000. Location, competition and economic development: local clusters in a global economy. Economic Development Quarterly 14(1): 15-34.

Ready, R.C., 1991. The value to Kentuckians of the Kentucky equine industry: a contingent valuation study. In: Understanding the impact of the equine industry in Kentucky and the Central Bluegrass. Center for Business and Economic Research, University of Kentucky, Lexington, KY, USA.

Stowe, C.J. and M. Rossano, 2013. Conducting a scientific survey of a state's equine population. Equine Disease Quarterly 22(3): 5.

Stringer, J., B. Thomas, B. Ammerman and A. Davis, 2013. Kentucky forestry economic impact report. Kentucky Woodlands Magazine 8(1): 8.

Part 3.

Economic opportunities

This part focuses on specific attempts to capture the economic opportunities presented by recent growth in the equine sector, particularly in terms of new activities and new ways of measuring them. The first chapter describes an attempt to professionalize the employment market in the equine sector in France, whilst the second describes the growth of the sector in that country, through the lens of new descriptions of what consists of, for example, equine tourism. This review shows that economic activities tied to horses occupies a much wider territory than has been heretofore acknowledged. The third chapter looks at how genomics, a very new but lucrative and rapidly growing economic activity might be expressed within the equine sector. All give good examples of how the sector is not only growing, but providing new opportunities for economic activity.

4. Jobs in the French equine sector

C. Cordilhac* and C. Abellan
Equi-ressources, IFCE, 61310 Le Pin au Haras, France; claire.cordilhac@ifce.fr

Abstract

French equine industry is built around 52,000 companies (breeders, races trainers, ridding schools, etc.). Most of them are localized in the North West area and generate about 180,000 jobs.

Keywords: job, equine industry, employee, statistics

Introduction

Almost 100 professions have been identified in the French equine industry (show, betting, breeding, sport, teaching, transports, food...). Some of them cannot be identified easily, such as horses ridding clothes solders.

Built around four sectors (breeding, races, sport/pastimes and meat), the equine industry achieved in 2012 a turnover of Euros 14,5 billions[1], out of which 10 billion came from the horse racing sector. The functions exercised, the working conditions, the career evolutions vary according to the sector concerned.

Equiressources, an institution in the service of employment in the equine industry

The equine industry has its own institution specialized in equine jobs: Equi-ressources. It was founded in 2007 in partnership with many institutions of the industry (Horse and Riding French Institute, Employment Center, The France's horse industry economic cluster 'Hippolia' and Basse-Normandie region). Equi-ressources aims to regulate supply and demand in terms of employment met. Equi-ressources is also a 'lab' that aims to observe and analyze the employment within the equine market. Finally, Equi-ressources stands as the institution that can provide young public with information, advice and orientation, which is key role when it is about to choose one's training or start one's career in the equine industry

Jobs in the equine industry

In France, the equine industry generated about 179,392 jobs in 2012 (57,301 as main activity), which represents 0.7% of total employment in France. Since 2013, a new counting system has been used by the Economic and social observatory of the French Horse Industry (OESC) similar to the one used by the International Labor Organization (ILO). This new counting system highlights an important part of multi-activity in the equine industry. The definition of the 'equine scope of jobs' is built on the 'closeness with the horse': Direct jobs (126,436; and 42,401 as main activity) represent the jobs 'in contact' with the animal (breeder, trainer, groom, rider, farrier, etc.). Indirect jobs (52,956; 14,900 as main activity) represent the jobs of those who play a role in the economic sector but have no direct link with the horse (sale of food for horses, administrative staff of an institution dedicated to the horse sector, etc.).

---

[1] Les observatoires économiques régionaux, les chiffres de la filière équine française, 2013. Available at: http://tinyurl.com/pmz6rbs.

A very large number of professions are part of the equine industry: most of them require basic skills and others require high skills (engineers, project managers, researchers, etc.). Hence, the scope of opportunities is very large and covers many fields such as agriculture, sports, government, trade, leisure and even art!

Despite the international financial crisis and the increasing rate of unemployment in general in France, the employment rate in the equine sector is dynamic. The profile of the population in the industry is rather young and feminized, half of the employees are in precarious situations and the turnover is relatively high.

Employee population in the equine industry

Equine industry's employees are young, the average age is 31 years old, or 9.5 years less than the national average age for employees of all sectors[2],[3]. The median age is 28 years old and the group of 18-22 years old represents a quarter of employees. Most of them are aged from 18 to 65 years old (almost 63%).

In 2012, 48% of the employees recorded by the MSA concerned unlimited-term contracts, 29% of jobs are related to limited-duration contracts, 14% jobs as apprenticeships contracts and 9% are related to seasonal works. Since 2009, the number of unlimited-term contracts and apprenticeships contracts increased (respectively +4% and +1%) while temporary contracts and seasonal works decreased (respectively -3% and -2%).

The situation of women is slightly worse than that of men. Even if there are more women than men in this industry, fewer women have a permanent contract. Women on average have €150 lower pay than men. In the breeding sector the difference can reach €260.

30 trades in the equine industry

According to Equi-ressources – the French job centre specialized in the equine area – there are about 30 different professions in the equine industry, 7 of them overall accounted 70% of this sector: groom, trainer, instructor, training rider (for races), horse grooms, breeder's assistant and equestrian tourism guide (Figure 1).

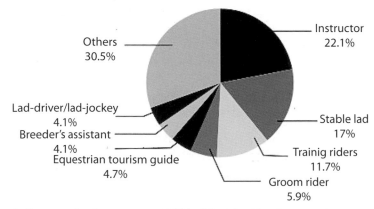

Figure 1. Seven professions represent 70% of the job offers (www.equiressources.fr).

---

[2] French mutual company specialized in farming.

[3] 'Evolution de l'âge moyen et de l'âge médian de la population', INSEE, 2012. Available at: http://tinyurl.com/q6hqbgz.

As it is often the case in a dynamic industry, tensions can arise: some of them are of a 'quantitative nature' (lack of applications to satisfy an offer (instructors, training riders, farriers), some are of a 'qualitative nature' (despite many candidates, employers do not find the skills they are looking for (professional equestrian riders, grooms, farriers for some specialties).

Conclusion

As a conclusion, the equine industry is a dynamic industry within the French agricultural sector, despite it represents a small part of the employment. However, the new counting system makes comparisons with previous years very difficult.

5. The equine business: the spectacular growth of a new equine segment market in France

G. Grefe and S. Pickel-Chevalier*

ESTHUA (Etudes Supérieures du Tourisme et de l'Hôtellerie de l'Université d'Angers), Université d'Angers, Laboratoire ESO-CNRS (Espaces et Sociétés) and Laboratoire GRAMEN, UMR 6590, 7 Allée F. Mitterrand, BP 40455, 49004 Angers, France; sylvine.chevalier@univ-angers.fr; gwenaelle.grefe@univ-angers.fr

Abstract

A social revolution in riding has created incredible growth in the equine-product market. This new equine economy is, in fact, characterized by the range of activities available (32 riding styles are listed by the French Equestrian Federation), by riders' needs (equipment for both riders and their horses including fences, water troughs, horse-boxes, etc.), by product ranges (from entry-level to luxury goods), but also by fashion which, thanks to the profile of today's horse-riders (predominantly female and young) has become a key part of the market. This enthusiasm has led to the development of businesses, the majority of which have appeared in the last thirty years in France (Antarès, CWD, GPA, Cheval Shop, EquipHorse, Fautras, etc.). While they have become key references in the sector, and account for nearly half of all jobs related to the equine sector in France (38,900 agricultural jobs compared to 32,800 non-agricultural jobs including business, veterinary care, farriers and public bodies), their characteristics remain largely unexplored.

Keywords: equine, business, growth, market, innovation

Introduction

Since the 1960s and 70s horse riding has become part of consumer and leisure society that triumph (Corbin, 2001; Dumazedier, 1962; MIT, 2005), to the point where it is now a commodity. The logic of taking riding-lessons, and as such, being initiated in the art of an elite culture (Franchet d'Espérey 2011; Roche, 2008), has been surpassed by that of hedonism (Digard, 2009; 2004). The French Equestrian Federation had 145,071 members in 1984, 434,980 in 2000 and the number now stands at 689,044 (2014)[1], with an estimate of around 2.2 million riders overall, including outdoor and casual riders who are not all club-members. The FFE is today the third-largest sports federation in France and the first female-dominated (Defrance, 2011; Tourre-Malen, 2006). However, this growth has been accompanied by a profound change in the way people ride (Chevalier and Dussart, 2002; Grefe and Pickel-Chevalier, 2015) coming from changes in rider profiles and their expectations. Seventy percent of French riders in 2014 were aged 18 or younger and 55% were 14 or under. In addition, 82% of all these riders were female[2]. Riding has also become part of a process of relative social diffusion concerning mostly the middle and upper-middle classes, as well as the upper classes (Terret, 2007; Tourre-Malen, 2009). This change of profiles and the fact that riding has moved into today's consumer society has created a diversification of needs and supplier to meet those needs. While clubs, which are anxious to retain their customers (Chevalier, 1998; Grefe and Pickel-Chevalier, 2015), strive to increase the range of activities on offer (show-jumping, dressage and cross-country, but also pony games, fun games, voltige, horse-ball, outdoor-riding, western-style riding, etc.), traditional practices are being replaced by a 'new baroque equestrian culture', meaning very heterogeneous practices (Digard, 2009: 16).

---

[1] http://www.ffe.com/journaliste/Publications/Statistiques.

[2] http://www.ffe.com/journaliste/Publications/Statistiques.

This social revolution in riding has created an incredible growth in the equine-product market. This new equine economy is, in fact, characterized by the range of activities available (32 riding styles are listed by the FFE), by riders' needs (equipment for both riders and their horses including fences, water troughs, horse-boxes, etc.), by product ranges (from entry-level to luxury goods), but also fashion which thanks to the profile of today's horse-riders, predominantly female and young, has become a key part of the market. This enthusiasm has led to the development of businesses, the majority of which have appeared in the last thirty years in France (Antarès, CWD, GPA, Cheval Shop, EquipHorse, Fautras, etc.). While they have become key actors in the sector, and account for nearly half of all jobs related to the equine sector in France (38,900 agricultural jobs compared to 32,800 non-agricultural jobs representing business, veterinary care, farriers and public bodies – IFCE, 2011), their characteristics remain largely unexplored.

The purpose of this study is to better understand those equine businesses. How they are characterized and where they are located across France? How do they respond to today's unprecedented diversification of demand, which has become much larger but also somewhat volatile? How do they also fit into a market characterized by globalization and increased competition? Do they borrow entrepreneurial logic from other sectors of the economy when setting their objectives and activities (searching for profitability and performance, adaptability or innovation), or do they manage to define a corporate culture which is specific to the equine industry by syncretism[3]? Indeed, can equine businesses invent their own logic between global entrepreneurial methods and particularities of equestrian needs and traditions? If the latter is true, does it allow new jobs requiring specific professional training to emerge, and which are unique to the equine sector?

Methodology

To answer these questions, we focus primarily on the growth of this new economy and its distribution across France. We determine which types of business characterize this highly specialized emerging sector, and where they are located around the country. We then focus on an analysis of how they have adapted to the equine sector in a market economy. Finally, we analyse the process by which those involved in the sector improve their professionalism.

Our methodology is based on:
- A study of businesses selling equine products, listed by the IFCE (French Institute for Horse and Riding) and the *Annuaire du Cheval*. The latter is a private directory published by a specialist magazine called *L'Eperon*. To be listed in this directory companies need to pay a fee. Nevertheless, its reputation in France has led the vast majority of companies, both large and small, to sign up. This approach will allow us to identify the different types of companies already in the market (small-businesses, chains, and large groups), their areas of activity and their distribution across the country.
- An interpretative analysis of interviews with the managers of 12 companies specializing in the market for equine products in order to understand, by combining and comparing their answers, what growth and progress strategies are employed by companies representative of the market.
- The study of how the Professional Degree in Marketing of Equine Products, at the University of Angers (ESTHUA[4]) on the Saumur campus, is run. This professional degree is unique in France in its focus and receives offers of placements and is visited by companies from around France looking for to recruit students. These graduates are highly employable. The analysis of how this training is run – the criteria for joining the course, the organization of its content, its statistics, its employment success rates, the sectors concerned and the location of these jobs –

---

[3] Synthesis of two or several cultural beliefs or practices from different origin, giving rise to new cultural forms.

[4] Esthua: Faculty of Tourism and Hospitality of the University of Angers.

will allow us to study the emergence of a new professional profile specialized in the marketing and sale of equine products.

An abundance of supply in response to changes in demand

The horse riding market had deeply changed, according to the recent evolution of the practices and representations of horse, that we called the 'contemporary revolution' (Pickel-Chevalier and Grefe, 2015). The sector is marked by a combination of social processes linking distribution, rejuvenation, feminization, anthropomorphism and hedonism (Digard, 2009). This phenomenon allowed the equine products and services to explode. The *Annuaire du Cheval 2013* listed 2,470 French and foreign companies present in the country. They are mostly young companies (mainly founded since the 1990s) and with varied structures. They are dominated by a plethora of small family-run businesses, punctuated by a few chains (Padd and Horsewood which merged early this decade, Horse Shop) and groups with a large market presence (Decathlon, Go Sport, Terres et Eaux).

These 2,470 listed companies include manufacturers, wholesalers and distributors, which are unequally divided between the following nine sectors of the industry: tack and saddlery (representing 730 companies, so 30% of the market); transport, infrastructure, food and health, but also equestrian tourism agencies, blacksmiths (producers of horse-shoes), firms specializing in consulting and services, and the arts and media including the specialist press.

While the market looks like it is a niche, given the fact that less than 4% of the French population regularly or occasionally go riding (2.2 million riders out of a total population of 65 million – INSEE, 2013), it has a structure which seems to be leading to a metamarket (Kotler and Lane Keller, 2009). By metamarket, we mean a market which gather multiple sub-markets such as transportation, food, health, tourism, etc. In fact, this net of apparently heterogeneous markets is characterized by a marketing relationship providing an ample supply of products linked to the initial purchase, that of a horse or hours of riding-lessons (Lambin and Moerloose, 2012). The activity of riding, but more importantly the purchase of a horse leads to an increase in need of equine products, today partly created by specialist companies using the model of mass distribution by combining the phenomena of fashion with an emotional tendency for anthropomorphism (for horses, but also for riders' equipment).

A specific analysis of the sector of tack and saddlery (the distributors of which may also be the manufacturers), which represents one third of all the equine product firms, allows us to observe trends in the market. The first observation to make is that if the large specialist distributors seem low in number (there are really only three players in the market, the first of which is Decathlon, the second being Go Sport, with Terre et Eaux coming third) they represent 52% of all tack shops (with 380 outlets). As such, even if their reputation is a subject of debate, such specialist shops represent just over half of tack sales outlets in France. We can't say that they behave like oligopolists, since almost half of the market is composed of small and independent companies. Beside, those large distributors are more specialized into entry-level products for leisure, whereas high-end products for competition are sold into smaller specialized shops. They testify about the inscription of riding growth into the more global development of sport of nature (Bessy and Mouton, 2004; Pickel-Chevalier, 2015).

An analysis of the location of tack shops in mainland France reveals that the whole country is covered, with every area having at least one saddlery. This reflects an equestrian activity spread across the country which nevertheless varies greatly from one region to the next. In fact the number of tack shops per area varies from 1 to 26 (Figure 1).

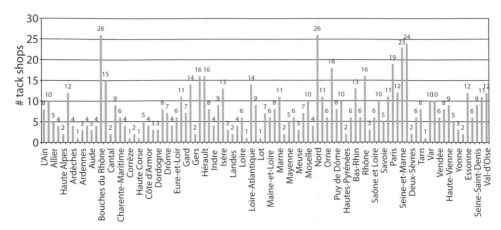

Figure 1. Distribution of tack shops per area in France (l'Annuaire du Cheval, 2013; design by S. Pickel-Chevalier in 2013).

Their geographical locations show important similarities with the locations of riding-centres, which themselves correspond to departments[5] of high population density (IFCE, 2011). Like riding-clubs and schools, the tack shops are therefore as close as possible to areas with a high concentration of riders, in the major cities (Grefe and Pickel-Chevalier, 2015). As such, only four departments of France have between 23 and 26 tack shops, namely the Bouches-du-Rhône in the South East (26 stores), the North (26 stores), the Yvelines (24 stores) and the Seine et Marne (23 stores) in the Paris region. The eight departments making up the Ile-de-France have a total of 116 outlets which represents 16% of all the whole French market. This high concentration corresponds to the structure of the French population: the Ile-de-France has 11.8 million people out of a national total of 65 million which is equivalent to 18% (INSEE, 2013). Riding is now an urban activity (Grefe and Pickel-Chevalier, 2015; IFCE, 2011) and tack shops take advantage of both an abundance of riders and a greater level of purchasing power than the national average. The Bouches-du-Rhône in the south of France (in the Provence-Alpes-Côte d'Azur region) is in the same situation as it is the third most populous area in France with nearly 2 million people, more than 850,000 of whom live in the city of Marseille (INSEE, 2013). The population of the Bouches-du-Rhône also has a higher income level than the national average.

This high purchasing power is not however present in the area which comes third in terms of the number tack shops, namely the north. However, it is the most populous region in France with over 2.5 million inhabitants (INSEE, 2013) of which more than 220,000 live in Lille. The high density of tack shops reflects the historical importance of the sporting mass-market retailing sector, due to the founding of the Decathlon chain in 1976 in the region. Today, the company, which has stores in 18 countries worldwide, has a high local concentration with 11 Decathlon stores distributing tack equipment and its own sub-brand, Fouganza, just for equestrian products. The area also has 3 Go Sport, 3 Padds and 3 Horsewood shops. Large stores therefore dominate the local market where the population does not need to have high purchasing power due to more competitive pricing.

Finally, the areas with the fewest number of tack shops corresponds roughly to the least populated regions in France, extending from the South (Ariège), South-Center (Tarn-et-Garonne, Lot, Haute-Loire, Corrèze) to the North-East (Ardennes) (Figure 2).

---

[5] France is made up of 95 counties called 'departements'.

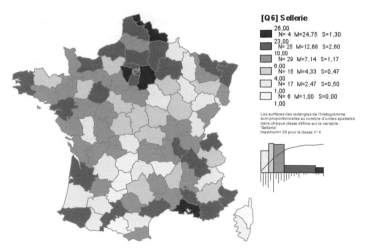

[Q 6] Sellerie

26,00
N= 4 M=24,75 S=1,30
23,00
N= 25 M=12,68 S=2,60
10,00
N= 29 M=7,14 S=1,17
6,00
N= 15 M=4,33 S=0,47
4,00
N= 17 M=2,47 S=0,50
1,00
N= 6 M=1,00 S=0,00
1,00

Les surfaces des rectangles de l'histogramme
sont proportionnelles au nombre d'unités spatiales
dans chaque classe définie sur la variable :
'Sellerie'
maximum= 29 pour la classe n° 4

Figure 2. The geographical location of tack shops (distributors) in mainland France: a spread which corresponds to population densities, urbanisation and purchasing power (design by S. Pickel-Chevalier in 2013).

As such, the location of tack shops in mainland France is more linked to population density, urbanization levels and the purchasing power of the population than to riding traditions or breeding centres. So, the two areas which are the most heavily involved in the equestrian agricultural sector (breeding, racing, equestrian sports), the Orne (headquarters of the National Stud Farm) and Maine-et-Loire (headquarters of the IFCE and Cadre Noir de Saumur) only have six outlets each, of which the majority are large specialist chain-stores.

This location of tack shops, determined by the location of their customers, thus reflects the changing pattern of equestrian activities, moving from traditional agricultural and rural bases to become a leisure pursuit favoured by the relatively wealthy urban population.

It is now important to understand how companies have managed to meet the demands of the new market besides their policy of geographical proximity.

Adapting the trade in equine products, involving the market economy

To analyze the sales and marketing strategies which have allowed companies in the sector to adapt to the 'revolution' of horse riding in France, marked by the combination of distribution, rejuvenation, feminization, anthropomorphism and hedonism (Pickel-Chevalier and Grefe, 2015), we conducted a survey of eleven companies which display significant dynamism in their diversified activities. Between January 2012 and January 2014, we interviewed:

- *Arnaud Lièvre*, sales director of the up-market tack specialist Butet. Based in Saumur, this saddlery founded by Frédéric Butet in 1987 was taken over in 2012 by Arnaud Butet and Olivier Perdrix. The Butet saddlery represents French tradition, specializing in equipment for sport competitions, more especially show jumping. Its influence is international. Butet saddles are selected specifically by lots of riders at the Cadre Noir de Saumur. The company employed 29 people in 2012.
- *Noémie Renard*, marketing director of the up-market tack specialist CWD[6]. Founded in 1998 in Nontron (Périgord) by Laurent Duray, the CWD saddlery specializes in innovation and research into high performance for those involved in competitions. It is the worldwide leader in the sale

---

[6] CWD is the name of the company.

of up-market saddles and is a supplier to a number of world champions, such as Kevin Staut. It produces 3,500 saddles each year, half of which are sold for export. The company is also known for its very aggressive marketing strategy.

- *Gaston Mercier*, director and founder of the up-market tack specialist Gaston Mercier. Founded in 1998 in Aveyron by Gaston Mercier, a former European (1987-1989) and French (1988) champion in endurance riding, the saddlery is famous for making up-market saddles for endurance and outdoor riding combining traditional leather-working and innovations (such as double offset stirrups – Innovation Award for 2013 at the Salon du Cheval de Paris). It has national and international representation. His son, Manuel Mercier, joined the company in 2008. The company therefore remains a family business and had fifteen employees in 2013.
- *Albert Joel*, founder and director of the boot-makers Botterie Saumur Albert Joel. Founded in 1989 in Saumur (Maine et Loire), the company specializes in up-market made-to-measure riding boots, considered a luxury product (€1,500 to €4,000 euros per pair). The company uses traditional French leather working techniques and employs four people. It sells its products through national and international outlets.
- *Carlos Biclet*, director of Cheval Shop. Founded in 2005 in Loire-Atlantique by Carlos Biclet and his wife Corinne Guesdon, Cheval Shop specializes in equipment for horses and riders in the mid-range market, with many unique products. The company which employed 23 people in 2013 has five stores (Nantes, Ile-de-France, le Mans, Haras de Jardy) and a soon to open shop in Lyon. The company is the third biggest tack shop (for horse and rider equipment) in France in terms of sales after Padd and Horsewood which have now merged. Cheval Shop also develops its own products (*Time Riders*). Their market is mostly in France, but the managers are looking to expand into Russia and Asia.
- *Jean-Philippe Lucas*, sales manager of the GPA Sport group. Founded in 1972, GPA helmets (Groupement pour la Protection Automobile) initially made protective headwear for Formula 1 car racing (famously worn by Alain Prost and Nelson Piquet) before their inventor Michet Finquel decided to transfer this technology to riding hats. GPA Sport was founded in 1998 (with its headquarters in Toulon). GPA riding-hats are now considered to be the world's leading equipment worn by champions around the world in every equestrian discipline, including racing.
- *Valentin Vielledent*, director and founder of the company Poney Matériel. Founded in 2003 in Valenciennes (North), this company specializes in pony games equipment for horses and riders. It is the only company licensed to provide material for pony games competitions in clubs in France. It is known for having created, in addition to specialized equipment, a new fashion in equestrian clothing by borrowing styles from team sports.
- *Hervé Hoffer*, director of the company AJC Nature selling equine care products. Founded in 2005 in Alsace by Mr. Hoffer and his wife, this family business specializes in natural care products for the well-being of horses (natural worm treatments, therapeutic care for rheumatism, emphysema, etc.) which are sold in the mid-range price bracket. The company has three employees and sells its products via mail order throughout France.
- *Christophe Lesourd*, director of the specialist travel agency Cavalier du Monde. Founded in 2005 in the Paris area, the company sells equestrian holidays on five continents. This is the newest and smallest – with just 2 employees – of the main travel agencies specializing in equestrian tourism in France (Rando-Cheval, Cheval d'Aventure and Caval and Go), and is known for its 'responsible' attitude, having been ATR[7] certified.
- *Sandra Arderius*, personal assistant to the editor of the specialist industry journal L'Eperon. Established in 1936, L'Eperon is one of the oldest companies in the sector. It has an excellent reputation and is considered the reference for breeding and competition in France. The company also owns the website www.cavadeos.com and the specialist bookshop Cavalivres. It employed 25 people in 2013.

---

[7] 'Agir pour un Tourisme Responsable' is a french label of sustainable tourism: 'Act for Responsible Tourism'. It is an AFNOR certification based on social, cultural and environmental norms, but also politic of natural resources preservation.

- *Claude Lux*, journalist and technical director of Randonner à Cheval, the only French magazine specializing in outdoor riding. The magazine was founded in 2005 by Jean-Michel Millecamps, and is sold in newsagents and via subscription. Claude Lux is also the author of around thirty technical books on outdoor horse-riding, as well as novels for children on the subject of horses.

The companies were selected to represent varied dynamic sectors of the industry (saddlers, tack shops, boot makers, the specialist press, equestrian tourism, equipment, care products and riding hats). They have varied levels of success, from retailers which dominate the market (CWD, Butet, Cheval Shop, etc.) to newly emerging companies (AJC Nature, Cavalier du Monde). Their positioning also highlights the diversity of the sector, from the up-market (Butet, CWD, bottier Albert Joël, GPA, Gaston Mercier, etc.) to mid-range products (Cheval Shop, Poney Matériel, AJC nature, Randonnée à Cheval.).

Semi-organised interviews were conducted between January 2012 and January 2014. These professionals have close personal links with the industry, either as enthusiastic riders themselves or married to highly-skilled riders (Frédéric Butet, Albert Joël). Only the two gentlemen who took over the Butet tack shop business (Arnaud Lièvres and Olivier Pedrix) come from other sectors but they have now also started riding.

Results

The initial findings, as to what these people feel is the current state of the equine business, all show they believe the market is maturing no matter what activity is involved. As such, 'Horse-riding around the world has gained significant public awareness and so the services on offer have developed accordingly, and the number of outlets too. This market has matured,' says C. Lesourd, director of the travel agency Cavalier du Monde. This maturation has also led to a more professional market, characterized by increased competition. 'Since my arrival, I have noticed a fairly intense competitive activity mainly driven by aggressive marketing and pricing policies,' explained A. Lièvres. This trend is of course reinforced by the financial crisis which all professionals are subject to, but has not suffered from heavy losses.

The maturation of the market is also illustrated by a better-informed customer, thanks to the many media outlets available (press, internet, etc.), but still remains under the influence of opinion makers, such as instructors champions and of course media. Fashion is therefore the new factor, completely upsetting the market by responding to the innovative demands of young, female customers. A combination of colours, rounded shapes, aesthetics and a fickle trend' has completely upset a market once dominated by techniques and an undeniable tradition of austerity. As such, A. Lièvre says, 'Amateur customers seem be much more under the influence of marketing than technicality itself, while the foreign market is more alert to quality and high end products from France.' Carlos Biclet, who has well-understood this evolution and incorporated it into his products says, 'I can already tell you that next spring's fashion will be green!'

To meet these new market characteristics, the companies surveyed are developing business strategies which, aside from their specificities in terms of products and positioning, are converging on a common policy based on the elements as discussed in the following sections.

Diversification of products and product ranges to fit in with social differentiation

As such, the tack shop Cheval Shop has increased its mid- to high-range products by importing brands from Germany and the United Kingdom which were unknown in the French market place a decade ago. The purpose was to create interest and stand out in the market by offering various stylish products not found in other tack shops. This desire to develop ranges and types of products, is particularly evident in large chain stores such as Decathlon which offers 370 products for horse-riders, 231 products for training horses and 395 products for horses at rest.

The *Padd* website offers 12,000 references covering all areas of horse-riding. Even companies recognized as market-leaders, such as *L'Eperon,* are today looking to diversify with the acquisition of the journal *Sports Equestres* targeting younger riders involved in competitions, but also by acquiring www.cavadeos.com and *cavalivres* aimed at the general public. The majority of businesses are preoccupied with offering the best value for money, helped by good market knowledge via the media (internet), but also uncertainty over the perseverance of riders. Riding in France is characterized by a high rate of people abandoning the sport, with an average of two thirds loyal riders and one third new-comers each year (FFE, 2012). The average time these new-comers ride for, before giving up, is one year and most of them are children (Grefe and Pickel-Chevalier, 2015) so parents are reluctant to invest large sums of money upfront.

Identification and accessibility in this competitive market

Companies choosing the high end of the market have a constant priority to maintain their position by insisting on French-made products. Their positioning is based on the recognition of certification and in particular that of a 'Living Heritage Company' already acquired by tack shops such as Butet, Hermès, Gaston Mercier of the boot-maker Albert Joël. The certification celebrates local production based around the preservation of traditional know-how. New types of certification are starting to appear such as 'Made in France', which particularly concerns Butet, or other more local versions such as 'Made in Aveyron' for the Gaston Mercier tack shop.

Ethics, also using a policy of social certification (the ATER certification – working for responsible tourism acquired by Cavalier du Monde; or 'Made in France' proving that jobs are being created in the country supported) or with an ecological objective such as 'organic produce' developed by AJC Nature.

Innovation which today concerns many companies in the sector. Fautras horse-boxes stand out because of its continuous innovations (including horse-boxes with U.S.-style doors, horse-boxes with diagonal stalls, horse-boxes for 3 and 4 animals). These innovations often come about as a result of technology transferred from more creative sectors of industry (automobiles, cycling, and running). In the same way, Michel Finquel imported technology he had invented for Formula 1 drivers and incorporated it into riding-hats 'in order to provide the same level of safety for his young daughter when out riding' explains Jean-Philippe Lucas. Springer Stirrups, meanwhile, introduced the technology of clip-in pedals, very popular in cycling, to their stirrups. Even companies which claim to be traditional are opening up to innovation so as to retain their attractive up-market position. As such, Gaston Mercier leather saddles have added an innovative system of double offset stirrups. Frédéric Butet created the Practice saddle with revolutionary ergonomics. Without saddle-flaps it helps with training and riding-development by promoting a better sitting-position and closer contact with the horse.

The search for performance and innovation which today characterizes the sector needs to include these three marketing factors which now dominate consumer society. The tack shop CWD focuses much of its communication on this aim. It is based on the combination of a search for *comfort and ergonomics* to which riders are sensitive when motivated by pleasure (lighter and more comfortable clothing and leather which is easier to clean), but also a need for safety. The fantasy of 'zero risk' and new regulations are pushing brands to develop stronger, safer products. GPA riding-hats have been awarded certification at both European (CE)[8] and U.S. (ASTM; SEI)[9] levels to provide an unprecedented level of safety for the rider by combining carbon, textalium, aluminium and lorica horse-armour. In addition, the products must also meet expectations in terms of design and fashion which concern the entry level Decathlon range as well as up-market models.

The company GPA Sport has managed to produce a very safe riding-hat which is relatively large, at the risk of turning away female customer, but which is a fashion accessory and an indicator of social standing despite costing between 400 and 700 euros. For this positioning, wearing a GPA riding-hat, just like riding with a Butet or CWD saddle and or Albert Joel boots, gives riders sense of belonging to a group – a sort of 'tribalism' (Maffessoli, 2000)– which is very strong in the equestrian world. 'My name hardly ever appears on my boots. The brand is represented by a simple stamped logo, but those who wear Joel Albert boots recognize them immediately,' confided Joel Albert. These up-market boots, like any luxury object (Rolex, Hermès), need to be discreet and are an outward sign of wealth only shared by those in the 'community'.

The logics of import-export

In a market which has become globalised, equine businesses, although largely dominated by a network of small and medium-sized businesses, all develop international relations. This is a sales strategy that works especially well for high-end products playing on an idea of luxury (Hermès, Butet, CWD, GPA). Nevertheless, companies specializing in mid-range products also hope to conquer the markets in emerging economies in which riding is becoming popular, like Asia (Cheval Shop). In fact, even if they are not all able to sell their products abroad, companies in the equine industry are almost always involved in the import of materials or products because of the logics of company-relocation, which still continues, despite the willingness of some companies to move back to France.

The integration of the important trends of feminization and anthropomorphism

Feminization, expressed mainly from the importance of fashion in the sector, has led directly to a transformation and diversification of products. While clothing was cut along strictly masculine lines until the 1990s (i.e. the domination of the 'Barbour' straight cut), by the millennium there was a profusion of more feminine accessories. Trousers, shirts and jackets all incorporated curved feminine cuts and were available in a range of colours. This feminisation was even visible in the style of saddles people were buying. 'I was the first to use colour in a saddle, and did so specifically for female customers, 20 years ago,' said Gaston Mercier, who produced endurance saddles made of red leather. Albert Joël also reveals he has recently added colours in his high-end boot ranges, including red, at the request of his female clientele who are looking for something different from the usual black or brown.

Anthropomorphism has led to the creation of various products designed for the well-being of horses according to criteria which come more often from conjecture than from actual knowledge about the needs of horse (numerous carpets, ribbons and blankets in assorted colours). 'I want a girth strap for my horse because it suits him well!' proclaimed one young rider and horse owner aged 14 before changing his mind and saying 'because it protects the horse when jumping...' (Grefe and Pickel-Chevalier, 2015).

---

[8] The CE (Communauté Européenne) mark, is a mandatory conformity marking for certain products sold within the European Economic Area (EEA) since 1985. The CE marking is also found on products sold outside the EEA that are manufactured in, or designed to be sold in, the EEA.

[9] ASTM and SEI are American norms of security that give the possibility to sell products in the USA.

Feminization, trends, the simultaneous search for a 'community spirit' and differentiation (which is specific to fashion) are leading to an exploitation of different riding activities in response to the needs of individuals looking for social recognition, expressed by a need for consumption that fits with the logic of community or neo-tribal identification (Maffesoli, 2000).

This desire to identify oneself with a world recreated by an activity (Cova, 1996) is widely used by shops in the equine sector with a range of equipment associated with each type of 'equestrian culture' as a sort of code that you belong. Thus, while the idea of 'jumping' suggests an atmosphere of 'competition or innovation' the idea of 'dressage' is more related to classicism and elegance. Endurance favours bright or fluorescent colours coming from more extreme sports (cross-country running, trail-riding, mountain biking), while pony games invented a concept inspired by team sports (bigger and more masculine clothing). Western riding helps to project participants into the myth of the 'American dream' by using cowboy-inspired tack (western saddles and bridles, shirts, jeans, long chaps with tassels, cowboy hats, etc). Finally, ethological riding ('natural horsemanship') also allows people to imagine the great U.S. or Australian 'outdoors', but via other methods (casual kit and fashion, an essential ethological halter... and a baseball cap in a 'horse friendly' approach).

The professionalization of tools to encourage sales

In the context of a globalised market, and one turned upside-down by very 'aggressive' marketing strategies, word-of-mouth, which was once the rule, is no longer enough. The trade in equine products now fits with more professional market logics, which requires tools which drive forward sales to be, similar to those in other sectors. Marketing has become a key tool for brand development, based on the creation of logos, slogans, and communication policies which may well be aggressive.

Carlos Biclet says, 'We chose black and red for our logo so as to be easily identifiable and to stand out in a crowded trade-fair, such as the *Salon de Paris*.' The sponsorship of successful sportspeople has been adopted by many professionals, such as CWD or Wintec, while others prefer to offer trophies (such as Cheval Shop) or decorate the jumps in their corporate colours (L'Eperon, CWD) during well-known competitions, as is the practice in other sports (motor sport, Olympic Games). A new marketing strategy adopted by Butet is aiming to set up a championship called the '*Butet Amateur Tour*'.

It is also becoming important to have people to influence public opinion, such as popular riding champions who are willing to be role models or even big-name stars of the riding world (Kevin Staut, Nicolas Tousaint, Isabel Werth, etc.), especially among younger riders. 'I am in constant contact with all the champions competing with the CWD team. I send them a text message after each win, regardless of the time difference if they are not in France. It is a fundamental relationship of trust and association between us,' explains N. Renard, marketing director for CWD.

Communication policies are also based on extensive use of social networking sites (Facebook, Twitter and forums) which are especially effective with most young riders, 70% of whom are 18 years old and under (FFE, 2013). They are in the process of developing both physical and psychological personalities (Dolto, 1988) and searching for identity recognition, which leads to mimicry. Websites are also becoming essential at every level of trade in equine products. From the biggest groups to the smallest private companies (AJC Nature, Cavalier du Monde, and independent tack shops) today they all have a website, the purpose of which is twofold:
- E-commerce (online sales) for both large and small businesses (Décathlon, Cheval Shop, Padd, Horsewood, Poney Matériel, AJC Nature, Cavalier du Monde, but also the specialist press such as l'Eperon or Randonner à Cheval).
- Communication for specialty brands that are only available via direct sale from the factory or in the consumer's home, particularly in the context of tailor-made products (Albert Joël, Butet,

CWD, Gaston Mercier) or through retailers (GPA, Fautras). Their sites, as a showcase for the brand, are nevertheless very 'well-designed' with plenty of detail and interactivity.

The internet can also be used as a tool for diversification, particularly in the case of L'Epéron, which opened a site offering information and interactive feedback which was more attractive for young people: www.cavadeos.com. You have to pay to register (but some information is available free of charge) and allows members access to over 20,000 articles and 3,000 videos related to the horse-riding world. 'We do not believe in the death of paper, but paper and digital media can exist side-by-side,' declared S. Arderius. The site is very popular and influential with over 100,000 visitors every month.

This professionalization of sales also results in a strengthening of direct and personalized relationships with customers. All the managers interviewed emphasized this requirement in relation to their clients, even when they finally commit to buy on the internet. Carlos Biclet confirms that despite the company having a very dynamic website, keeping and developing the physical shops seems indispensable in a context where the fear of 'fictitious' companies – which do not exist outside the 'internet browser' – is growing. Direct contact with the customer in a shop reassures them, he says. Trade-shows have the same role in allowing sellers to meet and get to know their customers, especially for companies selling mainly via the internet, such as AJC Nature.

'If we miss important shows such as Paris or l'Equita'Lyon our customers think we have gone out of business,' says H. Hoefer. New media does not necessarily destroy the direct relationship, but rather transforms it. 'It is essential to listen to customers, to be able to answer questions quickly by phone or email so as to reassure clients and acquire their loyalty,' says C. Lesourd (Cavalier du Monde). In this context, the number of tools to generate loyalty is growing. Copying the techniques of large retailers (Décathlon, Go Sport), specialist companies are now creating their own loyalty cards (Cheval Shop, Padd), sometimes combined with a system of sponsorship (Cheval Shop has introduced a system whereby a client with an internet account earns points for introducing a new customer). This loyalty process works especially well since it uses the logic of belonging and a community spirit which people look for in horse-riding, beyond simply joining a club, as proven by the success of national forums.

As such, the analysis of marketing strategies adopted by companies specializing in the equine trade leads us to observe the adoption of certain trends in the market economy, which is characterized by:
- the development of ranges (from the high-end specialized products via niche products to entry level products);
- adaptability and flexibility;
- differentiation;
- customisation;
- direct relations;
- accessibility.

However, these new strategies are marked by a syncretism with existing equestrian cultures. They do this by adapting to the cultures of the horse world, especially where the desire to maintain traditions remains a leitmotif (Butet, Devoucoux, Albert Joel), even when linked with innovation (CWD, Gaston Mercier, Forestier, Fautras, etc.).The ubiquity of equestrian culture can also be seen in the profile of business leaders who, with a few rare exceptions, start out as experienced riders. This quest for a combination of economic and entrepreneurial rationality and a passion for riding seem to characterize the professionalism of those in the industry, which today demands specialized training.

The professionalization of new faces in the equine product market

Professionalization of actors from both supply and demand also requires new recrutee profiles. Companies are now looking for qualified young people, coming from university courses, who are horse lovers. A passion for horses has become a necessary factor but is no longer enough to be a professional in the equine trade. If today's generation of entrepreneurs is still dominated by those who are self-made men and women, and are often inventors or visionaries (Fautras, GPA Sport, Frédéric Butet, Gaston Mercier, Carlos Biclet, etc.), they are now looking for young graduates combining both equestrian and business skills. In this context, many of them are choosing to take on trainees (for between 4 and 6 months) for a trial period and to test them before confirming their contracts. One way this is happening is in the innovation of the Professional Bachelor Degree[10] in the Marketing of Equine Products, created in 2002 at ESTHUA[11], which is the only specialist university course in this sector in France. This professional Degree is actually the third specialized in equine economy created in this faculty, after a Bachelor Degree Equestrian Sports and Tourism Management and a Bachelor Degree Riding-Centre Management. In 2006 had been created a fourth Degree at the level of Master: Sport and Tourism Management. The cursus are focus on several sports, but equestrian is a major one. Those creations of equestrian academic Degrees in ESTHUA testify about the needs of qualified staffs in the sector.

The Bachelor Degree of Marketing of Equine Products is well known by professionals who visit each week (1-3 companies per week) to present their businesses and select students for placements after conducting individual interviews. These business leaders come from all over France, but especially from the western part of the country and the Ile de France (Paris area) due to the ease of access to the university.

This recognition among professionals attests to the emergence today of a skills profile that really meets industry needs. 'I come back each year to look for future employees from the Professional Degree in the Marketing of Equine Products which corresponds exactly to our expectations. Cheval Shop has expanded thanks to this training,' says Carlos Biclet. To further understand this emerging profile, we will now focus primarily on how the training is organized and the criteria for students wanting to join the course.

As with any professional degree in France it is completed in one year, after two years of preparatory university study including internship. During that year it combines lessons on industry knowledge (the history of riding, knowledge of public bodies, equine taxation, equine law, etc.) and specialist lessons on business and communication (negotiation, marketing, communication, accounting, management, business start-ups, website creation and specialist English). Lectures given by professionals, interspersed with company visits, are held each week to allow students to fully discover the extent of the equine market, but also the responsibilities and the character traits of the invited company managers. These guests finish their lectures with individual interviews to identify students for placements.

The training also includes on-the-job experience through two training course a year, at the Salon du Cheval in Paris and in specific companies for a four months period at the end of the academic year.

This combination of theoretical knowledge, regular contact with professionals within the industry and internships, produces graduates with great employability prospects, which contribute to the

---

[10] Such professional degrees are in line with the general EU policy for higher education which a sequence of three years after high school to obtain a Bachelor, then five years to obtain a Masters and then eight years to obtain a PhD. All of this sequence of higher education degrees is also called 'Bologna process' and has been adopted by all EU member countries.

[11] Esthua: Faculty of Tourism and Hospitality of the University of Angers.

success of the programme. Its reputation is now confirmed, as evidenced by its use in recruitment across the country. Program leaders have seen, since 2010, how aspiring students make contact with the course co-ordinators as soon as they have finished their high-school diplomas in order to find out what the best training is to guarantee inclusion on the course. While in 2002 many candidates were often young people wishing to join the Degree on Riding-Centre Management but unable to qualify due to a lack of riding skills, we now see a high level of specialization straight out of high-school.

This emerging profile of a future specialist is characterized, like the rest of the sector in France, by a high proportion of women, a number even higher than among horse-riders. 90% of the students enrolled on the course are in fact girls – compared to 80% of riders (FFE, 2012). Eighty percent are on their first university course, which means that only 20% are repeating students. The group is therefore young (around 20 years of age). Unlike the recruitment for students for the Bachelor Degree on Riding-Centre Management, the students on the Marketing of Equine Products course do not come from agricultural and rural origins but are mostly from urban backgrounds.

75% of them come from marketing courses and their proficiency in English is one of the criteria for joining, as this skill is highly sought-after by professionals. The level of riding skills is less important than their management training, as they are likely to end up running riding-schools, but gallop level 5 (referred to as 'rider' level) is necessary, as is several years of riding experience and a passion for the sport. This feature is essential in the recruiting process because professionals come to the university looking for skills in both riding and business. A good knowledge of the sector is therefore vital.

In a study conducted 6 months after graduation[12], an in-depth analysis of the employment prospects from the 2011 class-intake revealed[13]:
- 58% were in employment (half of these found work directly after finishing the course);
- 27% were continuing their studies (Business School, Masters in Marketing;
- Communication, Masters Tourism and Leisure Sports);
- 10% were looking for employment;
- 5% were on a sabbatical year, usually having gone to an English-speaking country to improve their language skills.

We noticed that 73% of students were employed on permanent contracts. This particularly high rate comes after a six-month internship, serving as a long trial period, allowing the company to commit themselves without taking risks. 'After two trade-fairs and six months of internship we know the students and they know us. We therefore know if we can work together,' confided C. Biclet. Ninety one percent of these employees are in full-time jobs. Equally, their wages are in line with the average for their degree level: 71% are paid between 1000 and 1,600 euros net per month[14], compared to only 14% below average and 14% above average. This situation, which has been ongoing for several years, attests to a sector of industry which is part of the national economy and which, despite the financial crisis which started in 2008, has not seen a significant decrease in the number of new jobs created.

If we map the locations where they are employed (Figure 3), the class of 2011 are today working mainly in the western of France even though the students originally come from across the country. This phenomenon illustrates the logic of the professional network they are introduced to during

---

[12] 95% of the students has been graduated.

[13] The implementation of the precise inquiries of follow-up of the students is recent in the ESTHUA and are complex because the faculty welcomes 3,000 students. The results of 2011 were available in 2013. New results should be available in 2015 to deepen inquiries. But according to the coordinator of the Bachelor Degree who keep an active network with the old students, the statistic of 2011 are representative of the usual employment after graduation of the students.

[14] In 2012, the SMIC (Minimum Legal Salary) in France was of 1,118.36 euro a month. The students are usually paid when they start around 1,200/1,300 euro a month + profit-sharings.

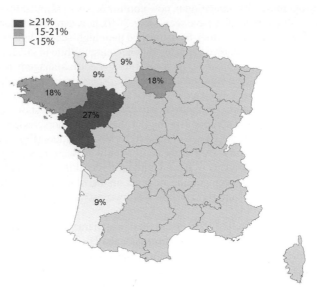

Figure 3. The employability of students from the class of 2011 (design by ESTHUA, University of Angers).

their studies, and which prevails over family logic. While students on the Degree in Riding-Centre Management, who are also recruited from across France, generally return to their hometowns where they have a company waiting or a family link, those on the Marketing of Equine Products course, and who often come from urban areas, create their professional network during their training. They are mostly employed in the western half of France, where the companies recruiting them are based.

Conclusion

The trade in equine products has matured by assimilating professional strategies and general market trends for consumer goods where, with the need to compete and stand out, two types of companies coexist: small and medium-sized family companies and large groups. If both are dynamic and successful, the large groups are upsetting the entrepreneurial model of the traditional industry. They oblige the small and medium-sized companies to assimilate the new tendencies of the economy, which are:
- flexibility and diversity;
- accessibility and non-stop communication, through an increased use of modern media (trade-fairs, Facebook, web-sites, newsletters, etc.);
- innovation;
- safety;
- product value and value for money.

This change has not however taken place by ousting the previous equestrian culture, but by a process of syncretism. The logic of a link, which characterizes the passion for horses – a sense of belonging to a 'community' where social ties prevail, leading to a type of tribalism – is a fundamental component of an sector 'where everyone knows each other and where we have always known everything,' according to many industry insiders. We are witnessing the emergence of a commercial equine economy. This borrows heavily from the global consumer market while maintaining its individuality associated with belonging to a 'cosmos', a community spirit – despite big divisions in the industry

– and a willingness to preserve traditions. Innovation is important but it doesn't sweep away the old equestrian culture. It modernizes and reworks it, rather than completely reinventing it.

This evolution nevertheless requires changes and heavy investment by small and medium-sized companies in a very competitive environment, or they risk disappearing from a dynamic market (but one limited to less than 4% of the French population), and one which is restructuring itself and producing transformed and diversified profiles. The quality and ethical positioning ('made in France') allows the high-end market to resist moving production overseas and instead invest in high technology, innovation and ethics[15] But the balance of forces is difficult to maintain; the relocation of production, especially for 'fashion' products where design is more important than quality, is widespread. The weakness of small and medium-sized companies is exacerbated by the good price/quality ratio of groups, such as Decathlon which specializes in innovations, and this is constantly increasing.

These developments require increased professionalization, including passion, but not only this. Beyond the generational divide – old school/new school – new job profiles are emerging. This is defined by new business skills and including the now very well-understood precepts of marketing. The analysis of professional training programmes (Marketing of Equine Products and Riding-Centre Management) which have been running for the last 10 years has allowed us to witness the emergence of a new profile of professionalism in the equine business. Breaking away from the previous logic of family and rural ties, these new employees define themselves, just like their customers, as young urban dwellers who love riding in a leisure context and do not necessarily have links to the traditional equine industry. They have the distinction of being associated with a logic of neo-tribalism formed around riding, while being able to stand back and adapt professionally given the diversity of cultures present in horse-riding.

The equine trade is therefore well ingrained in the global market for consumer goods while retaining its own peculiarities. Under pressure, but also stimulated by the requirements of a changing and diversified client group, it has resulted in a totally new equine economy leading to a metamarket and encouraging the confirmation of new job profiles. These are characterized by a dual insider/outsider relationship to the market and equestrian 'tribalism'.

References

Bessy, O. and M. Mouton, 2004. Du plein air au sport de nature. Nouvelles pratiques, nouveaux enjeux. In: Cahier Espaces 81 – Sports de nature, Évolutions de l'offre et de la demande, Editions Espaces tourisme & loisirs, Mai 2004, 17 pp.

Chevalier, V., 1998. Pratiques culturelles et carrières d'amateurs: le cas des parcours de cavaliers dans les clubs d'équitation. Sociétés contemporaines 29: 27-41.

Chevalier, V. and B. Dussart, 2002. De l'amateur au professionnel: le cas des pratiquants de l'équitation. L'Année sociologique 52: 459-476.

Corbin, A., 2001. L'avènement des loisirs, 1850-1960. Champs-Flammarion, Paris, France.

Cova, B., 1996. The postmodern explained to managers: implications for marketing. Business Horizon 21: 15-23.

Defrance, J., 2011. Sociologie du sport. La découverte, Paris, France.

Digard, J.P., 2004. Une histoire du cheval. Actes Sud, Paris, France.

Digard, J.P., 2009. Le cheval, un animal domestique au destin exceptionnel. In: Arts Equestres, Revue 303 Arts, recherche et créations, 12-19.

Dolto, F., 1988. La Cause des adolescents. éd. Robert Laffont, Paris, France.

Dumazedier, J., 1962. Vers une civilisation du loisir? Edition du Seuil, Paris, France.

---

[15] For example the companies Butet and Devaucoux have purchased a laser cutting tool giving better profitability when preparing leather parts, saving time and reducing waste off-cuts of material.

FFE, 2012. Dossier fidélité des licenciés 2013. CE Le Paddock. Available at: www.ffe.com/journaliste/publications/statistiques.

Franchet d'Espèrey, P. (ed.), 2011. L'équitation française. Le cadre noir de saumur et les écoles européennes, doctrines, traditions et perspectives. Lavauzelle Editions, Paris, France.

Grefe, G. and Pickel-Chevalier, S., 2015. De la transformation des établissements équestres en France lorsqu'ils intègrent la société des loisirs et de consommation. In: Pickel-Chevalier, S. and Evans, R. (eds) Cheval, tourisme et sociétés/Horse, tourism and societies. Mondes du Tourisme, Hors Série, Paris, France, pp. 136-149.

IFCE, 2011. Panorama économique de la filière équine. Les haras nationaux, Haras du Pin, France.

INSEE, 2013. Institut National de la Statistique et des Etudes Economiques. Available at: www.insee.fr/fr/regions.

Kotler, P. and K. Lane Keller, 2009. A framework for marketing management. Pearson Education, Upper Saddle River, NJ, USA, 384 pp.

Lambin, J.-J. and C. de Moerloose, 2012. Marketing stratégique et opérationnel du marketing à l'orientation de marché. Edition Dunod, Paris, France.

L'Annuaire du Cheval, 2013. Hors-série de l'Éperon. Cavadeos.com L'Eperon, Paris, France.

Maffesoli, M, 2000. Le temps des tribus: Le déclin de l'individualisme dans les sociétés postmodernes. Essai, Poche, Paris, France.

MIT, 2005. Tourisme 2. Les moments de lieux. Edition Belin, Paris, France.

Pickel-Chevalier, S., 2015, Can equestrian tourism be a solution for sustainable tourism development in France? Loisir et Société / Society and Leisure 38: 110-134.

Pickel-Chevalier, S. and Grefe, G., 2015. Le cheval réinventé par la société des loisirs en occident: une mythologie révolutionnée? (XVIII-XXIe siècle). In: Pickel-Chevalier, S. and Evans, R. (eds.) Cheval, Tourisme et Sociétés/Horse, Tourism and Societies. Mondes du Tourisme, Hors Série, Paris, France, pp. 26-49.Terret, T., 2007. Histoire du sport. Presses Universitaires de France, Paris, France.

Tourre-Malen, C., 2006. Femmes à cheval. La féminisation des sports et des loisirs équestres: une avancée? Edition Belin, Paris, France.

Tourre-Malen, C., 2009. Évolution des activités équestres et changement social en France à partir des années 1960. Le Mouvement Social 229: 41-59.

6. Initial approach to define the potential market of recent biotechnologies in the sport horse industry: the case of cloning

A. de Paula Reis

AgroParisTech; Current address: Ecole Nationale Vétérinaire d'Alfort, Department of Animal Production and Economy, 7, Av. du Général de Gaulle, 94704 Maisons Alfort, France; alline.reis@vet-alfort.fr

Abstract

The aim of this work was to design a tool to help stakeholders to understand the potential market of horse cloning (a disruptive technology) under current conditions. The work was performed in four phases: an extensive review of literature, observation of the life of a company and interviews with stakeholders, a numerical estimation of the potential market for cloning from the data published by FEI (for jumping, Endurance, Eventing and Dressage) and FIP (for Polo) and the proposal of a tool for quali-quantitative estimation of the potential market of a given region. In order to identify a responsible potential market for cloning we limited the candidate population to 0.3% of the professional horses. The opacity of the horse industry was a complicating factor in our study. However we were able to estimate the potential market for cloning at 191 to 267 candidates. The potential of Jumping, Endurance, Eventing and Dressage was 100, 31, 25 and 12 candidates respectively and the potential of Polo was 24-100 candidates. We considered that the candidate population was renewed in a turnover of 2-3 years. The quail-quantitative tool consisted on the combination of 5 main factors: the size of the horse population, the use of the horses within this population, the value of the best horses, the culture of the local stakeholders (which influences regulatory issues), and the knowledge of the horse market by the cloning actor. The results obtained in this work can guide the cloning actors in their strategic decisions and can also help breeders (and associations) to consider rational management of the employment of the technology in their breeding programs.

Keywords: cloning, horse, potential market

Introduction

Progress on biotechnologies of reproduction had been inserted in a process marked by progressive technology complexity, narrower target markets and more subjective benefits (the benefits of some recent biotechnologies cannot be easily measured by economical metrics).

To briefly illustrate this affirmation, we can compare Artificial Insemination (AI), Embryo Transfer (ET) and cloning the more recent technology of reproduction with an impact on the horse industry. These three technologies were all disruptive in their beginning because they had the potential to transform the traditional exploitation system and brought a new value and/or created a new market in the horse industry. For example, AI associated with adaptations of the sanitary laws contributed to protect the sanitary status of the herd (Margat, 2006), reducing the transmission of reproductive diseases. The commercial and sanitary benefits, the relative simplicity and the relatively low costs of AI contributed to improve its acceptation by the whole equine industry, except the Thoroughbred market that developed a different model of valorisation of stallions. Currently AI concerns more than 50% of the saddle bred broodmares in France and more than 90% in Germany (Reis, 2015).

ET was also a disruptive technology in the 70-80's. The concept of surrogate mares employed on ET made it possible to produce more than one foal/mare/year, to obtain offspring of active sport mares and of mares with poor uterine conditions. Undoubtedly, ET contributed to raise the economic value of mares by increasing their reproductive potential. However, the increased complexity and exigency of ET in terms of knowledge and resources compared to AI restrained the employment of

ET to females with high genetic status or animals with high affinitive value. Currently, the market of ET is restrained to less than 5% of the broodmares in France and in Germany (Reis, 2015). ET marked the transformation of the market of biotechnologies into narrower markets also influenced by subjective motivations.

In the case of cloning, ethical debates combined with the fear of loss of the genetic variability and the lack of regulation or prohibition reduced the willingness of breeders to adopt cloning. It unreasonably constrained its initial development. When we looked back at the history, we also observed passionate debates and prohibition of AI and ET by the breeders associations in the early, experimental, stage of development.

However cloning presented two new issues: it was conceived to be employed on a reduced number of individuals and its development must cope with the progression of the status of the horse to a pet animal (Olsen *et al.*, 2006) which imposed new ethical attitudes in the horse industry as discussed by Campbell (2013). The current opposition to cloning is probably linked to the lack of formal information about the technology, the employment, the target market, etc. as described by Myers *et al.* (2002) in the case of disruptive technologies.

In the last years our work was dedicated to the clarification of different aspects of cloning (Nakhla and Reis, 2012; Reis and Palmer, 2010; Reis *et al.*, 2012). Our work highlighted the need of extending the base of formal information in order to improve the quality of the debates and improve strategic positioning of different stakeholders in regard to the responsible management of cloning.

The aim of the present work was to supply initial information about the potential market for a responsible cloning activity. The arising estimation was not intended to define the exact cloning market but to supply stakeholders with a first quali-quantitative method of judgment for identification and estimation of the potential market.

Methodology

This work was a case study as described by Eisenhardt and Graebner (2007). The work was performed in four phases: the first phase was a deep review of the literature about the horse market and the management of innovative technologies; the second phase consisted in the observation of a cloning company and the interview of different stakeholders; the third phase was dedicated to study different segments of the horse market in regard to their numerical potential to generate candidates for cloning. In the fourth phase, we proposed a comprehensive tool to allow actors to identify countries / regions or disciplines with higher or lower potential for cloning from little initial information. This tool could be important to help an actor to structure his developmental strategy and could help studbook managers to evaluate whether or not the breed could be reasonably concerned by cloning and in which intensity (potential number of candidates).

It was important to understand the horse market and the mechanisms of disruptive innovations in order to develop the present study. The review of the literature performed in the first phase was extensive in order to embed some principles of the market estimation method proposed further. It included technical aspects (technical knowledge available, efficiency, etc.), management of innovation (pathways of development, crystallisation of the developmental model, innovative ecosystems), the economy of singularities (opaque markets, value of singularities, etc.) and the study of the horse market (overview, organisation, new tendencies, etc.).

The second phase of the study was performed simultaneously to the first phase and was aimed to understand the ecosystem (Agogué, 2012) in which cloning has been developed. It involved the observation of the activities of a cloning company for three years, visits to equestrian and scientific events in different countries and stakeholders' interviews. The observation of the company was employed to understand the difficulties of cloning development, to identify possible driving factors of this situation and to structure the semi-directive interviews of the stakeholders. The semi-directive method, discussed by Combeisse (2007) was not intended to quantify the answers but to identify the qualitative perception of cloning within different groups by means of the strong representations revealed. In total, 216 semi-directive interviews were performed (7 scientists, 5 cloning actors, 128 breeders, 21 veterinarians, 4 directors of breeders associations and 51 other horse users) distributed in France, Belgium, Brazil, Argentina and United Emirates. These countries were chosen because of their dynamic breeding activity and/or scientific expertise necessary to support a cloning activity. The stakeholders were identified by their activity in the horse industry.

In the third phase the different segments of horse industry were analysed in regard to their potential to generate candidates for cloning. The multiple methods of estimation of the horse population in different countries, breeds or sports compelled us to look for unique global sources for each parameter: global horse population and number of horses involved in equestrian activities that add value to horses. This methodological choice supplied information standardised enough to treat the different global regions in the same manner. Data from FAO (FAO, 2009) was selected to supply information about the world horse population. The analysis of the equestrian segments was restrained to the horses with an active inscription at the Fédération Equestre Internationale (FEI, (FEI, 2012)) (Jumping, Dressage, Eventing and Endurance) and the athletes with active inscription at the Federation of International Polo (FIP; (FIP, 2012)) and the rankings of studbooks of the World Breeding Federation for Sport Horses (WBFSH (WBFSH, 2011)) because these are the more complete sources of official information about equestrian activities on a global level. The objective of this approach was to propose a quantitative estimation for this market based on the professional equine population presenting high economical interest. In order to quantify the potential population of candidates for cloning we proposed a threshold. The aim of this threshold was to ensure that the population identified in our study would answer relevant commercial and technical criteria: an outstanding own performance, a high visibility of the individuals at the athletic level and reduced risk for genetic progress or variability. The threshold was set on 0.3% of the performers inscribed in the official competitions, all genders included, and was based on an interpretation of studies about methods of genetic selection (Arnason and Van Vleck, 2000; Koenen and Aldridge, 2002), about the impact of cloning on the genetic progress of a population (Ricard and Dubois, 2006) and about the market practices of breeders towards cloning (Reis and Palmer, 2010). To complete the phase of numerical identification, we compared the data issued from FEI (2012) for official equestrian activity and from WBFSH (2011) for breeding rankings in order to estimate the geographical development of breeding and sport activities for the different disciplines. Polo does not have an equivalent to WBFSH. The main breeding activity of Polo horses is performed in Argentina. The data from FIP (2012) was exploited only to identify the main countries developing Polo sportive activities.

In the fourth phase we integrated the numerical, social and commercial information obtained in the previous phases in order to propose a tool to help players to identify countries with low or high primary potential for cloning. This tool consists in a simple dynamic approach to guide players on their developmental strategy from a few parameters.

Results

We reviewed the literature in different sectors: methods on sociology, economic theories, horse market in several countries, ethics in the horse industry, genetic selection and cloning technology. The First four sectors were necessary to embed our questions and develop an appropriate method of investigation and to interpret the results. The technical literature supplied first elements to identify scientists working on cloning and to supply elements of discussion with the stakeholders interviewed. The main technical aspects to retain for the present work were: the low number of scientists working on this technology, the very initial stage of development, which explains at least partly the low technical efficiency and the high costs of production and the importance of the first movers on the funding of the research.

In the second phase the interviews with scientists highlighted a strong competition within the scientists for the private funds which was reducing exchange within players and slowing the technical progress and further transfer to the field. The scarcity and high logistical costs to obtain oocytes were important barriers for further technical improvement.

The commercial players were developing their activities in a restrained market (less than 10 clones produced per year, all producers together). The high prices, lack of information about the market, market potential and technical state of the art stimulated the arrival of new entrants and induced strong unstructured competition. As a consequence, players experienced low strategic capacity. Only players with good knowledge/contacts on the local horse market were accepted by the stakeholders as legitimate spokespersons.

The breeders of different countries showed different expectations about cloning. Indeed, in South America the technology was perceived as a tool for genetic progress and for new market opportunities increasing the willingness to employ the technology. The two main obstacles to the acceptation in South America were the high prices and, in some cases, religious concerns. In the European countries, cloning was globally perceived by breeders as an unethical technology. However, the genetic and financial values of the best horses in Europe were compatible with cloning and allowed the initial development of the technology. In United Emirates the barriers were the specialty of the local business model, the fact that a local production was not yet possible, regulation and, in a less extent, religious concerns. In both cases, the main motivations for cloning were to produce high genetic value cloned stallions or broodmares, to extend life of high affinity horses. The lack of technical and market information were frightening factors related to animal welfare and reasonable employment of the technology.

Veterinarians were identified as important stakeholders. In South America they were an important source of information contributing to the breeder's decisions. In Europe the majority of them preferred to keep some distance from this activity. Only few European veterinarians were disposed to prescribe cloning to their clients.

Riders and other users in South America perceived cloning as an intriguing technological advance while the Europeans showed many *a priori* but their position could evolve with some clarification of the technology (the main issues reported were health and the regulation concerning the register and use of clones). These stakeholders were not inclined to employ the cloning technology. However, in all the countries visited the riders interviewed would accept to ride a clone or an offspring of the clone of famous horses.

The observation of the life of a cloning company highlighted that the breeder's decision of cloning a horse can take several months to some years. This decision was influenced by four main factors: social acceptance, information about the technology, regulations and the career progression of the candidate horse during the decision making process. In the field of social acceptation we could identify family, friends, collaborations and even religious convictions as important factors. The important information included: observation of the clones and their offspring and answers to technical questions: what is cloning? Is it genetic manipulation? What about the health of the clones? And the longevity? The regulatory questions involved official authorisation for competition (i.e. FEI) and studbooks registration of clones and offspring. Finally, the progression of the performance or health and ageing of the candidate horse during the decisive process were also influencing the decision. Regulatory questions and individual progression appeared to have important impact on the subsequent economic exploitation of clones.

Despite the rational interest of reducing the interval between the performance of the candidate and the birth of the descendants of the clone (to improve the genetic gain) (Ricard and Dubois, 2006), we did not observe any customer decision to clone a horse in the early stage of the professional career of the candidate. The customers preferred to observe the global career potential of the horse before deciding to clone.

The analysis of the world horse population and equestrian official segments showed that the world horse population was estimated to 40,022,222 horses distributed in Latin America (47.05%); North America (24.69%), Europe (11.3%), Africa (11.3%) and Oceania (1.02%) (FAO, 2009). As the whole horse population is not involved in commercial equestrian activities we analysed certain sports segments in order to evaluate the part of the horse population involved in these activities. Data from FEI and FIP supplied us with valuable information.

From FEI (2012) we were able to estimate the number of the sport horses participating in official competitions and the number of potential candidates for cloning (as defined in the material and methods) Jumping: 33,316 and 100, Endurance: 10,189 and 31, Eventing: 8,173 and 25 and Dressage: 3,902 and 12, respectively (Figure 1).

The estimation of the Polo market was more complicated and demanded a different analysis. The analysis was not precise but could be of value for a first study about the cloning market potential on this discipline. Indeed, the information available at the Polo Association was oriented to the identification of the polo players and their respective competition levels and did not identify the horses, even if they are an important factor of success for Polo. In order to maintain a numerical approach our estimation of the number of horses involved in Polo official competitions was based on the number of players inscribed at FIP and took into account an important rule of Polo: each player must prepare 4 horses / match. Thus we identified 11,406 players registered at FIP which represented 45,624 horses at the international level. The estimation of the potential candidates for cloning was based on the number of players inscribed at the highest level (Handicap 10). Only 24 players were inscribed in this category. Our knowledge of the international market of Polo Horses, the breeding system and the interviews with stakeholders in Argentina allowed us to estimate that at least one horse of each Handicap 10 player could be a potential candidate for cloning. In some cases, the 4 horses player could be candidates for cloning. With this methodology involving a numerical approach and field expertise, we were able to estimate the potential market for Polo Horses within 24 and 96 horses. We estimated that this methodology was satisfactory because the candidate population represented 0.05 to 0.2% of the horses participating in Polo official competitions and is within the limits of the threshold proposed in our methodology.

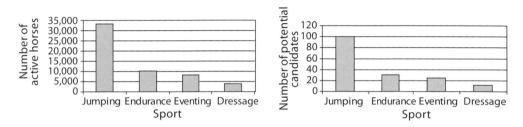

Figure 1. Number of active horses in official competitions of FEI (left) and potential of jumping, endurance, eventing and dressage to provide candidates for cloning (right) (adapted from FEI, 2012).

The geographical analysis of the activities described in the material and methods highlighted a complex geographical dynamic linked to the international market of horses and the international activity of the main riders. It prevented a clear geographical definition of the market. Nevertheless, we noticed a tendency to the development of important breeding/exploitation axes specific for each sport (Table 1).

These axes were an important observation because cloning is a breeding biotechnology dependent on adult cells (animals with economic value, whose performance is one of the main influencing factors). If breeding and equestrian activities are not developed in the same country, the cloning actor must take it into account for his strategy of development.

From the results of the 3 initial phases we identified the 5 main parameters influencing the potential market for cloning. From the interviews we identified 2 social parameters: culture and knowledge of the local horse industry by the cloning actor. The local culture was a driving force of the regulatory aspects. When local culture was favourable to cloning, the studbooks were more receptive to the cloning technology in their breeding systems. In order to propose a simple tool, we proposed the 'Culture' as the main social parameter including regulatory questions. The second social parameter of importance was the level of knowledge of the market by the cloning actor. The more the actor was close to the market the more he could have access to the breeders and to discuss cloning issues.

From the analysis of the horse population and horse industry we identified the 3 other important parameters of the tool: the horse population, the use and the value of the horses. Indeed, none of these parameters appeared to be relevant alone. However, the combination of them supplied valuable information about the potential market for cloning. Taking into account the threshold of 0.3% of the population as potential candidate for cloning, as discussed in the methodology, larger populations of horses should show bigger potential for cloning. However, the type of activity was an important determinant of the value of the horses and their potential to become a candidate for cloning. In the occidental societies of the beginning of the 2000's, breeding and sports add more marketable value to a horse through the genetic market or the competition prices than leisure activities or ploughing.

Table 1. Main axes of production and exploitation of sport horses (adapted from FEI, 2012; FIP, 2012; WBFSH, 2011).

Sport	Main producers	Main users on high level of performance
Jumping	The Netherlands; France, Germany, Belgium	France, the Netherlands, Germany, UK, Sweden, Swiss, Ireland USA, Brazil Australia
Dressage	Denmark, Germany, the Netherlands, Spain, Portugal	USA Germany, the Netherlands, UK Australia East Europe
Eventing	France, Ireland, Germany	UK Australia, New Zealand USA Germany
Endurance	France, Argentina, Uruguay	Middle East
Polo	Argentina	Argentina UK USA Australia

In addition, because of the high prices of cloning (from €133,000 to €200,000) we proposed that the more valued horses of a population must attain values at least similar to the prices of cloning in order to support cloning activity.

In Figure 2 we represented the parameters that a stakeholder should consider when evaluating the market potential of a region in order to build his strategy.

The results of this work can help a cloning actor to adapt his strategy according to his business model: market segment; identification of candidates or sale of the right to clone or to adapt his business model to the market of a given region. If the actor is also a producer of clones, he must add to this tool the ease to obtain horse oocytes locally. We did not add this parameter in this work because it is an extensive question that was already discussed in a precedent study on management of production (Nakhla and Reis, 2012).

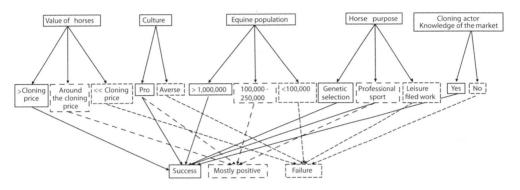

Figure 2. The complex combination of factors influencing the horse cloning market.

Discussion

The aim of this work was to supply initial information to help the stakeholders of the sport horse industry to build their strategy in regards to cloning: responsible acceptation of cloning activity and responsible investment on a cloning production/commercialisation activity. The lack of formal information imposed the construction of a bank of rational information in order to nourish reasonable debates about the subject instead of the current passionate existing ones.

We adopted the genetic argument to define the limits of a threshold of responsible cloning because we assumed that genetic variability and genetic progress are closely related to management of reproduction and the main function of clones is to be introduced in reproductive systems and to produce offspring. We assumed that the threshold chosen in our work (0.3% of the sport horses participating in official competitions) respected the rules of selection of sires and dams (2% of males and 20% of females of a given population) proposed by Koenen and Aldridge (2002) and Arnason and Van Vleck (2000) and did not represent a danger for genetic progress as evaluated by Ricard and Dubois (2006). Moreover, the threshold proposed in our work represented 1 candidate for cloning within 208,449 horses in the global equine population. Good practices of management of the reproductive career of the clone, as well as it must be done for every other sire or dam would allow avoiding risks of reduction of genetic variability, as explained by Gautschi et al. (2003) in a study about the genetic variability loss in a population of birds.

This quantitative estimation of a responsible potential market for cloning was necessary to help breeders to decide whether they can or not clone a candidate horse and could guide breeders associations in the rational management of the acceptation of cloning in their breeding systems.

Despite this important quantitative information, the interviews brought to light that the cloning market could not be estimated by traditional economic metrics and that the customer of the cloning technology is strategic. These two characteristics explained the complexity of the task of defining the potential market of cloning in a given region.

Conventional methods for market estimation were not useful to estimate the market of cloning in agreement with the observation of Bower and Christensen (1995) in other disruptive innovations because the conventional methods employ objective parameters and standard methodologies conceived for situations where the main environment is known. Horse cloning is an innovation developed in the horse industry. It brought together a double weakness of the existent information: the low level of knowledge about the horse industry (Couzy, 2008) which is the environment where cloning must be developed and the low level of knowledge about the environment of cloning itself.

In order to overcome the difficulty imposed by the lack of information and to propose a simple tool to help the estimation of the potential of a region to support a cloning market (Figure 2) it was necessary to understand some mechanisms of the horse industry including: the motivations of the stakeholders taking part in the horse industry and the strategy of consumption regarding cloning.

The interviews performed in our work corroborated with the research recently developed by the French National Studs HN (2009) and Couzy (2008) indicating that a noticeable driving force of the horse industry was the multiple subjective motivations of owning a horse. The researchers highlighted that the parameters influencing the purchase of a horse include several subjective factors like the comfort of riding, the behaviour / character, the breed, the level of athletic potential and the health status of the horse but also the own satisfaction or the satisfaction of a member of the family and even the acquisition of some societal status. This model of social value had been described by Karpik (2007) as the economy of singularities in which the estimation of value is driven by the search for 'the best' and this parameter can vary according to the person.

In our research the stakeholders interviewed agreed that only very good horses should be cloned. However, the definition of 'a very good' horse was variable within different users. We also observed a certain impact of the culture on the conception of a very good horse. Gradey (2008) explained that the value judgement in the economy of singularities is the fruit of a process of social construction. Studies in different sectors highlighted factors composing the value of a singular product. For example the quality of a painting, also a singular object, is composed of a complex subjective system, including the sensations transmitted by the product to the customer, its rarity, the network of the artist (Robertson, 2005) and the influence of the promoting gallery on the market (Sagot-Duvaroux et al., 1992). It could explain our findings.

The first elements of value of a candidate for cloning revealed by our work were the performance, the genetic potential, the public visibility and personal experiences (emotions provided by the horse). We observed that the most known horses of a discipline were more likely to be cited as desirable candidates showing the importance of the social value of a horse interfering with the potential for cloning.

The low objectivity to appreciate the value was probably the reason why the stakeholders interviewed were more open to discuss cloning issues (interest, risks, etc) with players having a deeper knowledge of the horse market. These players were supposed to have enough knowledge to decode the aspirations of customers, perform a good evaluation of the value of horses and identify the horses with potential for cloning (acting like prescribers as described by Karpik, 2007).

This was the reason why in our decisional tool (Figure 1), we proposed that knowledge of the horse industry is a factor of success for a cloning actor. As we observed that cloning a horse is not the result of a blind decision. In the current stage of the technology, the cloning actor must be able to clarify the breeders about the technology and the candidate's potential.

It contributed to understand the second main characteristic of the cloning industry: the customer (the breeder) has an active behaviour and his decision can evolve over time. Gallego and van Ryzin (1994) described the concept of the active customer as a recent consummation habit strongly linked to the development of online markets. We were not able to identify if this behaviour was exclusive of cloning as an innovative technology or if the horse stakeholders are mostly active customers.

Conclusion

In summary, the method of observation of the cloning market developed in this work supplied first formal information about the cloning market potential. It highlighted that the reasonable estimation of the market potential must be evaluated within the population (before identifying individuals). This work allowed us to determine a threshold for responsible cloning and roughly estimate the quantitative potential market for horse cloning. We also identified potential geographical poles of development of the market according to the sport disciplines. In addition we identified the active profile of breeders and the qualitative factors influencing the value of a horse in regard to cloning. Breeders, breeders associations and cloning actors can use this information to better understand the cloning market, the limits, and social aspects and to engage a rational dialog concerning acceptance and management of the technology.

References

Agogué, M., 2012. Modéliser l'effet des biais cognitifs sur les dynamiques industrielles. Innovation orpheline et architecte de l'inconnu. Thesis, Ecole Nationale Supérieure des Mines de Paris, Paris, France, 253 pp.

Arnason, T. and L.D. Van Vleck, 2000. Genetic improvement of the horse. In: Bowling, A.T. and A. Ruvinsky (eds.) The genetics of the horse. CABI Publishing, New York, NY, USA, 527 pp.

Bower, J.L. and C.M. Christensen, 1995. Disruptive technologies: catching the wave. Harvard Business Review, January-February: 43-53.

Campbel, M., 2013. When does use become abuse in equestrian sport? Equine Veterinary Education 25: 489-492.

Combeisse, J.-C., 2007. La méthode en sociologie. La Découverte, Paris, France, 128 pp.

Couzy, C., 2008. Un marché complexe pour un produit singulier. In: Institut Français du Cheval et de l'Equitation (eds) Proceedings of 2nd Journée du réseau REFErences, IFCE, Saumur, France.

Eisenhardt K.M. and M.E. Graebner, 2007. Theory building from cases: opportunities and challenges. Academy of Management Journal 50: 25-32.

FAO – Food and Agriculture Organization, 2009. FAOSTAT – live animals. United Nations. Available at: http://tinyurl.com/pqhcj8u.

FEI – Fédération Equestre Internationale, 2012. Rankings. Available at: http://www.fei.org/fei/about-fei/publications/fei-annual-report/2012.

FIP – Federation of International Polo, 2012. Polo athletes. Available at: www.fippolo.com/playing-countries.

Gadrey J., 2008. Le bon, le beau et le grand: entre culture et marché, les singularités. Revue Française de Sociologie 49: 379-389.

Gallego, G. and G. van Ryzin, 1994. Optimal dynamic pricing of inventories with stochastic demand over finite horizons. Management Science 40(8): 999-1020.

Gautschi, B., B. Müller, B. Schmid and J.A. Shykoff, 2003. Effective number of breeders and maintenance of genetic diversity in the captive beraded vulture population. Heredity 91: 9-16.

7. 'Great recession' impacts on the equine industry in the United Kingdom

C. Brigden*, S. Metcalfe, S. Mulford, L. Whitfield and S. Penrice

*Myerscough College, St Michaels Rd, Bilsborrow, Preston, PR3 0RY, United Kingdom;
cbrigden@myerscough.ac.uk*

Abstract

Britain's economy has been challenged since the 'great recession' of 2008-2009, with most risk to leisure industries. This research explored effects of the recession on aspects of the equine industry. The objectives were to investigate the recession's impact on: (1) equine businesses' marketing strategies; (2) affiliated British Showjumping (BS) participation; and (3) market value of horses. Equine business owners (n=133) completed an online questionnaire. BS participation was analysed using registration and competition entry data. BS competitor (n=144) and venue (n=32) questionnaires were used to explore experiences of the recession's effect. Horse values were compared by sampling advertisements of three groups of typical horse types. Interviews (n=3) were conducted with elite equine sales companies. Vendors and purchasers (n=74) completed a questionnaire investigating sales experiences. Associations within survey responses were explored using Chi-squared test of association. Some business owners (45%) felt the recession had affected business. Service providers were less affected than businesses that sell products or are directly involved with horses. Fewer small businesses (39.29%) reported recession effects than large businesses (63.19%). A new marketing method was the most common change to marketing strategies. Over 98% of businesses had attracted new and retained previous customers since 2008. BS membership numbers fluctuated, but the number of competitions and the number of entries per member decreased. Fuel price and the recession's impact were significantly associated ($P<0.001$) in influencing a competitor's decision to enter competitions. Riders who reduced the number of competitions entered, were significantly associated with reduction in the amount of training ($P<0.05$). Venues agreed that fuel prices had the greatest influence on entries, but despite 66% noting entry reduction few altered marketing or pricing strategies. Median non-elite horse prices reduced between 2007, 2009 and 2011 (£7,500, £6,500 and £6,250); approaching significance ($P=0.063$). Vendors and purchasers (73%) perceived a reduction in horse prices and related this to a reduction in disposable income. Elite sales data reflected a reduction in average prices in 2009, but subsequent increases in 2011. The leisure and non-essential aspects appear to be more vulnerable to recession. The equine industry should respond strategically to future recessions to ensure financial security.

Keywords: recession, horse, industry, price, consumer

Introduction

The equine industry in the UK is vast, containing an estimated one million horses, generating around seven billion pounds per annum and employing 220,000-270,000 people (British Horse Industry Confederation, 2009). Equestrianism is the eighth most popular sporting activity in England according to participation numbers, attracting in the region of 350,000 weekly participants (Sport England, 2013) and 3.5 million riders within a year (British Equestrian Trade Association, 2011). The equine industry can be described as being more varied than other sectors within agriculture or the leisure industry. Routes for consumer expenditure within the industry are manifold, ranging from the purchase of horses and related goods, to the recruitment of services such as livery, veterinary and farriery care, and the provision of sport-related opportunities, such as competitions and training. The 'great recession' of 2008-2009 posed a significant threat to the economic security of many industries, with leisure industries like the equine industry at potentially greater risk as cautious consumers are forced to rationalise the use of lowered disposable incomes.

questionnaire established information about the nature of participation, such as level of competition and professional / amateur status, before asking whether the respondent had modified his or her participation due to the recession and further details regarding this. Competitors were also asked for their opinions on the value of membership and prize money. Venues' questionnaires explored the business's success throughout the recession, for example whether a reduction in entries has been seen and their supposed reasons for these. The questionnaire then collected information regarding the venue's business approach to recession, such as whether increased marketing had been used, diversification or altered fees. Questionnaires were completed by 114 competitors and 32 venues.

Market value of horse survey

Advertisements posted within the UK's leading weekly equine periodical (*Horse and Hound Magazine*) were sampled for three groups of typical horse types in 2007, 2009 and 2011. The groups were 153-170 cm General Purpose, Dressage and Showjumping horses. These types were selected due to their popularity with British riders. Selected horses were standardised with regard to age (mature, 8-15 years), competition level (medium affiliated level or equivalent), experience and location. For each year, 20 adverts were randomly sampled for each horse type, half in mid-July and half in mid-November to take into consideration seasonal effects. Advertised price was collected for each advertisement sampled.

Interviews with three 'elite' equine sales companies or studs were conducted, which allowed collation of actual sale price data from the elite end of the market. Prior to the interviews, the participants were requested to supply data regarding their highest, average and lowest sales prices for in 2007, 2009 and 2011. The interview then explored the participants' opinions regarding the effect of the recession on sales prices.

Questionnaires were designed to investigate consumer and purchaser experiences of horse sales, such as the number of horses purchased or sold, and whether prices were thought to have increased or decreased. Recruitment of participants was promoted through the use of Social Media. Questionnaires were completed by 74 horse vendors and purchasers.

Statistical analysis

Questionnaire responses were exported into Microsoft Excel and appropriately coded where necessary. Statistical analyses were conducted using Mintab 15™ statistical analysis software. Questionnaire responses were analysed using Chi Square Test for Association to explore possible associations between categorical answers, with the null hypothesis being that that the categories in the two variables are independent (Dytham, 1999). For the analysis of the business responses, businesses were pooled into three categories (Table 2); 'direct', which incorporated businesses involving the care and training of horses directly (e.g.: livery yards, equestrian centres, studs); 'services', which provide services to horse owners (e.g. farriers, vets, physiotherapists); and 'products', which manufacture and market goods (e.g. feed suppliers, saddlers, tack shops).

Advertised horse prices did not conform to normal distribution according to an Anderson-Darling test (AD=2.813, P<0.005). Prices were therefore compared between years using Kruskal-Wallis test, with the null hypothesis being that all samples are taken from populations with the same median (Dytham, 1999). Kruskall-Wallis test converts raw data to ranks therefore eliminating the influence of any extreme data points (Dytham, 1999). Average prices for each type of horse (General Purpose, Dressage and Showjumping), horse genders and geographical locations were also compared using Kruskal-Wallis tests. P<0.05 was deemed as the critical value for significance in all statistical analyses.

Results

Equine business survey

Recession effects on business were felt by 45% of respondents. Fewer service providers (28.57%) felt the effects of recession, compared with businesses providing products (42.86%) or direct equine (53.97%) businesses. This association approached significance ($\chi^2_2=5.176$, $P=0.075$).

Effects of recession were felt by 39.29% of small businesses (employing 1-5 staff), 43.75% of businesses employing 6-10 staff, 57.14% of businesses employing 11-15 staff and 63.16% of large businesses (employing 16+ staff), but there were no significant associations between size and recession effects ($\chi^2_3=4.481$, $P=0.214$). Businesses who felt the effects of the recession were more likely to make changes to their marketing strategy ($\chi^2_2=84.574$, $P<0.001$). Where changes were made to the marketing strategy, owners were given an open question to describe the changes. Changes described were then categorised and are shown in Table 3. 81% of businesses reported that the altered marketing strategy was effective in attracting new customers, which was irrespective of the business type or size. Overall, 98% of businesses answered that they had attracted new customers and retained previous ones since the recession.

British showjumping participation

Total memberships fluctuated between the years, but the number of competition entries demonstrated a continual decline (Figure 1). This equated to an average reduction of 3.64 in the number of competitions entered per member per year.

Owners' perceptions of the effect of the economic downturn were not associated with numerous factors; whether horse was kept at home or livery, geographical location, whether prize money or affiliation was considered sufficient, level of competition and number of horses owned ($P>0.05$).

Fuel prices and the effect of the recession were significantly associated in influencing a rider's decision to enter competitions ($\chi^2_3=15.937$, $P<0.001$). 54% of riders stated that the recession and increasing fuel costs had influenced their competition schedule. 3.51% stated that only the recession and 27.19% reported that only fuel price increase influenced competition attendance. 14.91% reported that neither the recession nor increasing fuel costs influenced schedules.

Feeling the effect of the recession was significantly associated with a reduction in the amount of training attended ($\chi^2_3=11.039$, $P<0.05$). 38.60% had reduced the amount of training attended, of which 77% had felt the effects of the recession. 84.36% of riders felt that affiliation was worth it,

Table 3. Alterations to marketing strategies implemented by equine businesses in the UK in response to the recession (n=133).

Implemented alteration to marketing strategy	Percentage of businesses utilising altered marketing method
Increased spending on advertising	21%
Diversification	9%
Reducing costs	6%
Reduced spending on advertising	13%
Introduced new marketing method	51%

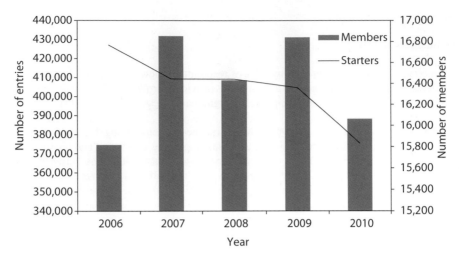

Figure 1. Total number of British Showjumping members between 2006 and 2010 in comparison to the number of show entries (starters).

but of those considering it affiliation unworthy, there was an association with also feeling that the prize money was not worth it (χ^2_3=8.673, P<0.05).

A reduction in BS entries was reported by 66% of venues, but this was not associated with feeling the effects of recession. The likelihood of venues noticing the effect of the recession was not associated with many of the recorded factors; whether cost of shows affected the number held, whether entries had reduced, whether fees had been changed, whether customer incentives were used, whether strategies were changed, whether the venue was used for non-equine purposes, whether the business had been diversified and whether Facebook, websites, texts, posters or emails were used for marketing (P>0.05). Websites remain the most popular form of marketing utilised (Table 4), but only 25% of venues altered their marketing strategy in response to the recession.

Venues who saw effects of recession were more likely see a reduction in entries for other competitions (χ^2_5=6.5045, P<0.05). Most venues (87.5%) felt that fuel prices had influenced business.

Market value of horses

A reduction in horse prices between 2005 and 2010 was noticed by 73%, although some respondents differentiated between types of horses. For example, 'Quality horses have held their price, mediocre horses are very cheap'; 'Top quality horses have maintained value, other have decreased'. 95%

Table 4. Marketing strategies implemented by British Showjumping venues (n=32).

Marketing strategy utilised	Proportion of venues utilising approach
Websites	100%
Facebook	25%
Emails	72%
Text messages	31%
Magazine advertising	3%
Poster advertisements	19%

stated that the recession had impacted on horse price. Respondents ranked reduced expendable income at most significant contributor to horse value reductions, followed by increased costs in bedding and feed.

From sampled advertisements, there was a trend for median prices to be reduced (2007: £7,500, 2009: £6,500, 2011: £6,250) (H_2=5.52; P=0.063). Prices were not affected by geographical location (H_9=12.85; P=0.169) or horse gender (H_1_0.58; P=0.445). Dressage horses (median £9,375) were significantly more expensive than Showjumpers (median £7,725) and General Purpose horses (median £4,375) (H_1=87.74; P<0.001). Data from sales prices recorded by elite horse breeders or sales companies are recorded in Table 5 and illustrate a general decline in 2009, but subsequent increase in 2011.

When asked to comment on whether the recession had affected some sectors of the industry more than others, interviewees had strong views, for example the owner of one of the warmblood studs responded:

I think that the sector affected most have been the mid-range priced horses and those not specifically targeted at a discipline, in other words the general riding horses. People in my experience are happy to pay for quality and will buy from reputable breeders. There are a good number of buyers in the market who go out with a budget in excess of £10,000, whose fortunes desires for the right horse are not overly affected by economic changes. This is the market we are targeting our sales.

Discussion

Customers within the equine industry express their preferences and priorities through their level of expenditure and repeated business for particular goods and services. It is of no surprise that service providers were less affected by the recession, since these usually provide services considered essential by horse owners, such as farriery or veterinary treatments. Expenditures that customers feel are luxuries and therefore dispensable, such as competition entries or training, will inevitably be reduced first when effects of recession are being felt. This pattern was seen in the reduction in competition entries and training made by British Showjumpers. The convincing majority of general

Table 5. Top and average sales prices recorded by horse sales companies (n=3) in 2007, 2009 and 2011. Total numbers of sales per year were not recorded.

Type of horses sold	Year	Top sales price (£)	Average sales price (£)
Hanoverians (mixed age)	2007	6,000	4,000
	2009	12,000	5,000
	2011	10,000	7,000
Trakehners (mixed age)	2007	8,500	5,000
	2009	12,800	7,239
	2011	26,200	10,737
Thoroughbred foals	2007	493,500	38,368.05
	2009	420,000	34,351.80
	2011	472,500	37,674
Thoroughbred yearlings	2007	2,625,000	67,677.75
	2009	735,000	54,462.45
	2011	1,785,000	66,509.10
Thoroughbred mares	2007	3,570,000	95,933.25
	2009	1,785,000	55,506.15
	2011	2,520,000	73,652.25

businesses were able to retain and attract new custom throughout and beyond the recession, suggesting a successful strategic approach. Businesses that were not able to withstand the recession are, of course, not represented within this sample, warranting further analysis. Claire Williams, Executive Director and Secretary of BETA, reflected that there was an increase in the number of businesses resigning from BETA in 2008 and 2009, although the percentage doing so due to closure had actually dipped. Resignations over this period were attributed to a slow-down in trade down and related cost cutting, as well as a competing trade fair which led to an increase in overseas members resigning their membership (Williams, C. Personal Communication, 21 March 2014). This illustrates how businesses rationalise their spending.

The influence of the size of businesses on how recession was perceived was likely to be an interplay between other characteristics. Small businesses appeared to be more resilient to the effects of recession, but these included many of the service providers, and therefore it is likely that the nature of business may be more influential than the size of business.

Businesses and organisations should review their customers' priorities with greater attention during times of recession. The level of overall satisfaction of the British Showjumping Association members was reassuring; with nearly 85% feeling it was worth affiliating. This result is potentially biased since the sampling strategy only included members of BS and therefore riders who feel affiliation is sufficiently invaluable to cause them not to join the association remained unrepresented here. A likely contributor towards the feelings of poor value for affiliation was prize money; riders dissatisfied with prize money were more likely to be dissatisfied with affiliation.

The influence of fuel prices is likely to impact most strongly on those businesses either reliant on customers' travel or those involving substantial travel to provide services or source goods. A model of the impact of fuel price alterations on driving behaviour based on data collected by the German Mobility Panel indicated that fuel-price elasticity is marginally higher for weekday travel, leading the authors to suggest that motorists are more responsive to higher fuel prices on non-work (more flexible) days (Frondel and Vance, 2010). It is of no surprise then that both venues and competitors felt that participation in showjumping reduced as a result of the fuel price rises. This is shown in particular by the reduction in the average number of entries per member per year, which suggests that even when riders feel it is worth paying for affiliation, they are choosing which competitions to enter more carefully.

Marketing strategies within the equine industry in the UK present an opportunity for development. 51% of general equine businesses implemented a new marketing strategy, such as a new way of advertising or changed pricing strategy, and reported beneficial effects on their customer base. This compares with only 25% of BS venues. The latter also demonstrate a rather outdated approach to marketing, favouring the use of traditional media such as websites and emails. The use of social media, despite its wide integration in other areas of business, was very limited. Approaches to expenditure on advertising were mixed, with 21% of businesses increasing and 9% decreasing spending. As Kotler and Caslione (2009) argued cuts to marketing should be considered carefully in times of recession, with resourceful companies having the potential to increase customer engagement and marketplace supremacy. It is clear that the ability of the majority of equine businesses to maintain customer numbers must be, in part, attributable to the finding that 91% did not decrease marketing spending.

Low prices at the bottom of the horse market continue to present concern to the animal welfare authorities since these are strongly linked to threats to welfare. The World Horse Welfare estimated that there are thousands of equines at risk of abandonment in the UK (Owers, 2014), a trend mirrored in other countries, such as the US and Ireland (Leadon, 2012; Stull, 2012). This concern was also voiced by the respondents to the questionnaire, for example, 'I think people selling horses for minimal amounts, like £1000 and giving them away is just creating a massive problem. The authorities have no idea the amount of horses and ponies out there that are not passported and with no regulations on livery yards, owners and dealers, it is creating many abandoned and mistreated animals'. Cunneen and Dana (2010) described the high sensitivity of horse markets to the economy, particularly during recession. The clear dichotomy between the lower end of the market and the elite, Thoroughbred market where top prices reach several million pounds, appears to have become more pronounced throughout the recession. This was illustrated by price trends within the elite sales sampled where price drops during the recession in 2009 seemed to be reversed by 2011. The non-elite horse prices, in comparison, continued to decline in 2011. Successful breeders highlight that horses must be produced strategically, with a clear purpose intended in order to achieve success, particularly during recession.

Conclusion

Examination of the impact of the great recession on the equine industry in the UK has highlighted some key strengths, in particular a trend for businesses to adapt their strategic approach to marketing to maintain market position. The leisure and non-essential aspect of the market appears to be more vulnerable to effects of recession. The sample size for each conducted survey was fairly limited and this study should therefore be considered as a basis for further development of research in this field. The fragmentation of the equine industry in the UK makes a comprehensive evaluation of the economic status of the whole industry challenging, but this preliminary work has highlighted that comparisons between fragments of the industry could be beneficial in identifying and sharing areas of good practice.

Members of the equine industry need to respond strategically to future recessions to ensure financial security. It appears that the most important member of the industry, the horse itself, may be at greatest risk from economic instability and this is a cause that the industry must attend to with immediate concern.

References

British Horse Industry Confederation, 2009. Size and scope of equine sector. BHIC, London, United Kingdom.
British Equestrian Trade Association, 2011. National Equestrian Survey structural report. BETA, Weatherby, United Kingdom.
Cunneen, H. and L.P. Dana, 2010. Actions and the New Zealand horse industry: What happens when global recession hits? International Journal of Business and Globalisation 5(3): 297-303.
Dytham, C., 1999. Choosing and using statistics. Blackwell Science Ltd, Oxford, United Kingdom, 320 pp.
Frondel, M. and C. Vance, 2010. Driving for fun? Comparing the effect of fuel prices on weekday and weekend fuel consumption. Energy Economics 32(1): 102-109.
Kotler, P. and J.A. Caslione, 2009. How marketers can respond to recession and turbulence. Journal of Customer Behaviour 8(2): 187-191.
Leadon, D.P., 2012. Unwanted and slaughter horses: a European and Irish perspective. Animal Frontiers 2(3): 72-75.
Owers, R., 2014. Responsible breeding and horse ownership: do you need to breed? Proceedings of the 22nd National Equine Forum, 6th March. London, United Kingdom.
Sport England, 2012. How we play – the habits of community sport. London, United Kingdom.
Sport England, 2013. Active people survey 7, October 2012-October 2013. London, United Kingdom.
Stull, C.L., 2012. The journey to slaughter for North American horses. Animal Frontiers 2(3): 68-71.

8. Equine entrepreneur's well-being

T. Thuneberg* and T. Mustonen
HAMK University of Applied Sciences, Degree Programme in Agricultural and Rural Industries, Mustialantie 105, 31310 Mustiala, Finland; terhi.thuneberg@hamk.fi

Abstract

Horse businesses such as breeding, riding activities and trotter training, are very labour-intensive, and physical enterprises. The duties of an entrepreneur are various, and include many different kinds of know-how: in addition to horse care, an entrepreneur should manage, for example, economic planning and business administration. There is a risk that the challenge of the work and the workload with long working hours can cause problems with well-being. For arm workers, long-lasting stress can lead to burnout as well as other physical or mental disorders and illnesses. Overall, mental health problems in farm workers have increased rapidly over the last years and are one of the major reasons for the premature retirement of the farm workers. HAMK University of Applied Sciences delivered a survey to equine entrepreneurs in 2011. Its aim was to find out how entrepreneurs themselves determine their well-being and work strain. The results indicated that Finnish equine entrepreneurs feel well, in general, but they face many challenges in their work which could have an impact on their well-being and coping. This type of entrepreneurship gives a certain liberty in organizing the job, but the flip side of the freedom is horses, which need 24 hours responsibility. Nevertheless, 66% of the respondents regarded their quality of life as good. Partners, family and friends offer important support for the entrepreneur. The network of other entrepreneurs is also a significant factor in maintaining their well-being and the management of businesses.

Keywords: equine business, entrepreneurship, well-being, health, coping

Introduction

The number of farms has decreased recently in Finland, in part because of the unstable economy. Some of those who have practiced traditional farming and quit farming have turned to equine enterprises. In addition to base production (agriculture and horticulture), one third of Finnish farms practices 'other businesses' (Information Centre of the Ministry of Agriculture and Forestry 2011). Diversification, entrepreneurialism and specialization are a big challenge for any rural entrepreneur. Producing a different kind of services increases the work load, which may lead to poor organization and being in a constant hurry. More and more strain is focused on the entrepreneur. According to Paasivaara (2009), good welfare consists of balancing work and leisure time.

Equine entrepreneur's weekdays are full of various physical and labour-intensive tasks, no matter which activity the entrepreneur practices: breeding, riding activities or trotter training. In addition to horse care, an entrepreneur should keep economic management and business administration under her/his control. Equine enterprises are usually small, and the entrepreneur her/himself is the manager. They are the ones who make contact directly with the customers. In that case, leadership is mostly self-leadership (Luukkala, 2011). Nevertheless, an entrepreneur must take care of animals all the time – whether there is demand for the services or not.

Motives on becoming an entrepreneur are individual and there are several reasons to start business. Economic motives are not uppermost in horse business – seeking to satisfy one's own dream is often the strongest motive. When the employment situation is bad, self-employment may also be the only way to find a job without having to move away. Seeking profit and risk assessment is nevertheless dwarfed by the preferred lifestyle (Heinonen and Järvinen, 1996).

Material and methods

The aim of the study was to find out how entrepreneurs themselves determine their well-being and work strain. In the winter 2011-2012 HAMK University of Applied Sciences targeted an e-survey to 196 equine entrepreneurs, who had participated in the national updating training during years 2010-2011. The questions of the survey consisted of multiple choice questions, and frequency and importance statements based on the Likert scale, but some open ended questions, too. The response rate was 33%. The majority of the respondents (78%) were full-time entrepreneurs offering, in most cases, horse riding or boarding services.

Results

Three fourths of the respondents were between 30 and 49 years old, which is slightly less than average age of farmers (50.7 years) according to Information Centre of the Ministry of Agriculture and Forestry (2011). An average respondent had almost 20 horses and 20 hectares of arable land, while the mean value of arable land on farms is 38.9 hectares in Finland. Work load has increased in 60% of cases during the entrepreneur's career, which varied from 1 to 27 years (average was 8 years). When asked to estimate the real work demand in man-years, entrepreneurs had difficulties to determine it – concept of 1800 hours per year wasn't clear. Half of the respondents reported that they work ten hours or more per day. The capacity for work reduced during the entrepreneur's career, and almost one fifth were unsatisfied and worried about their health (Figure 1). At least some health problems could be avoided with rationalization: by the use of proper tools and devices in every day routines. And further, health problems may be a reason for dissatisfaction with ability to manage the enterprise.

Generally, the work load was found quite strenuous. Respondents evaluated their work in five-stepped scale (from light=1 to extremely strenuous=5); half of them defined their work as at least quite strenuous mentally (mean 3.4), and two thirds physically (mean 3.7). Physical and mental tiredness is relatively common (Figure 2); over 75% of the respondents felt both the physical and mental tiredness quite often or sometimes. The most common way to keep coping, in general, was physical exercise, but also free time without horses and horse-related people were emphasized in the responses. Liberty to self-organize the duties is seen as positive, but on the reverse side of the freedom is horses, which are a constant responsibility. Nevertheless, 78% regarded the quality of life good or extremely good.

Full-time entrepreneurs had 16 days' vacation per year, on average, while the law-based vacation for the agricultural entrepreneur is 26 days (if included in the Finnish farm entrepreneurs' pension scheme). Different possibilities to have a substitute include using municipal or private substitute services, or relatives, friends and neighbours for helping out, which was the most popular choice

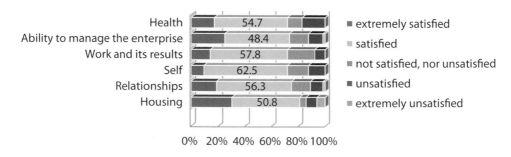

Figure 1. An equine entrepreneur's satisfaction to different life sectors.

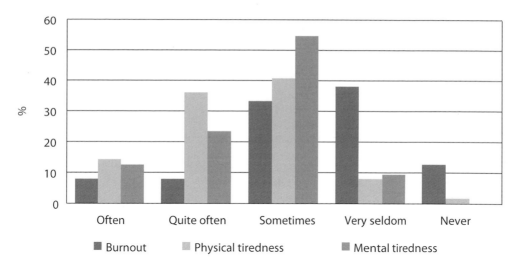

Figure 2. An entrepreneur's feelings about tiredness and burnout.

(38%) among the respondents. Majority of the entrepreneurs used also outsourcing services, like bookkeeping and machine contracting (including for example cultivation and pasture repairing) to get help for everyday life.

Conclusion

Finnish equine entrepreneurs consider their state of well-being to be good, in general. To maintain the health, an entrepreneur should take advantage of technical developments, such as feeders, which simplify stable routines and are quite rarely used in stables. Partners, family and friends form an important support network for the equine entrepreneur. A versatile network of other entrepreneurs is also a significant factor in maintaining their well-being and the good management of their enterprises.

References

Heinonen, S. and A. Järvinen, 1996. Maaseutuyrittäjän opas (*in Finnish*). Gummerus Kirjapaino Oy, Jyväskylä, Finalnd, 210 pp.

Information Centre of the Ministry of Agriculture and Forestry, 2011. Maatalouslaskenta 2010 (*in Finnish*), Available at: http://stat.luke.fi/e-lehti-2011-06-29/index.html.

Luukkala, J., 2011. Jaksaa, Jaksaa, Jaksaa – työhyvinvointitaitojen kirja (*in Finnish*). Kustannusosakeyhtiö Tammi. Kariston kirjapaino Oy, Hämeenlinna, Finland. 287 pp.

Paasivaara, L., 2009. Työnsä kokoinen ihminen (*in Finnish*). Kustannusosakeyhtiö Tammi. Kariston Kirjapaino Oy, Hämeenlinna, Finland, 132 pp.

9. How can horse business professionals adapt to new consumer demands in equine services?

S. Pussinen* and T. Thuneberg

Häme University of Applied Sciences, Degree Programme in Agricultural and Rural Industries, Mustialantie 105, 31310 Mustiala, Finland; sirpa.pussinen@hamk.fi

Abstract

The equine industry in Finland has shown considerable growth during the last decades. There are currently 75,000 horses in Finland, and the equine industry employs 15,000 people. Yet there are few studies concerning market conditions in the equine business. Approximately 3,000 stables provide business services. Häme University of Applied Sciences carried out a small-scale survey among entrepreneurs during spring 2013. The survey included questions about demand, profitability and plans to enlarge or reduce the size of these businesses. The entrepreneurs evaluated demand and profitability from the perspective of their own enterprises, both at the time and for the next five years. Most of the respondents came to the conclusion that demand would increase rather than decrease. In particular, entrepreneurs who offer riding and livery services or horse tourism foresee increasing demand, whereas they do not expect profitability to develop as positively. Nevertheless, almost half of the respondents have plans to enlarge their business. Even though the entrepreneurs are quite optimistic about the future, there is still a need for them to respond to changes in the market. Horse businesses are often regarded as a way of life rather than as a mere economic enterprise. Finding new customers, business planning and marketing communications must constantly be emphasised.

Keywords: equine business, entrepreneurship, demand, profitability

Introduction

There are 75,000 horses in Finland and approximately 3,000 full-time or part-time enterprises in the horse sector (Hippolis ry et al., 2014). There are both small-scale businesses with only a few horses, and large-scale businesses with 70 horses or more. Equine businesses primarily engage in breeding, trotter training, riding services and provision of livery stables. According to Pussinen and Thuneberg (2010), on average there are 26 horses per riding or trotter training stable and the turnover for the stables ranges from about 100,000 to 150,000 euros per year in Finland. Another study concerning the European horse sector obtained as a result, the monthly stabling cost per horse was at the most 6,000-9,000 euros per year (Liljestolpe, 2009). In Finland, there are 14.6 horses per 1000 inhabitants (Hippolis ry, 2010).

Häme University of Applied Sciences has made a study of the prospects of the equine business sector in Finland. The study concerns the Finnish equine industry and its economy. The aim of the study was to see the big picture, to see the whole equine industry and not only riding services or the trotting sector. The final report is titled 'The prospect of horse business in Finland', and it is published in Finnish as an e-report in the HAMK series of publications. The report's main results are described in this article. The orientation of the report is more practical than academic because the aim was to find out topical information for further projects and to analyse entrepreneurs' feelings related to difficulties in world economy in the spring of 2013. The report was a part of the 'Equine Businesses into Top Condition' education project (2010-2014) in which the main task was to organise further education for entrepreneurs. The project offered information about management and business operations, stable construction, environmental know-how and horse-breeding. The objectives were to increase entrepreneurs' competences and improve the profitability and the well-being of both the entrepreneurs and horses. The project was supported by the national, partly EU-financed Rural Development Programme for Mainland Finland (2007-2013) programmes.

One weakness of the Finnish equine industry is that there is no single common registry for all the enterprises, so accurate data are not readily available. The Finnish Trotting and Breeding Association have a detailed register for the horses in Finland. In addition to that they have information about 160 licensed trotting trainers. The other national association is the Equestrian Federation of Finland: 340 riding schools, leisure riding stables and private stables. These registers do not, however, give enough data about profitability, turnover, costs or customers – i.e. there is no basis for comparison – and this formed the background for the study. Answers were sought for questions concerning how does the equine industry in Finland as a whole see its economic prospects now and in the near future, how do entrepreneurs estimate the present state of the horse sector and what the motives to continue horse business are?

Material and methods

A survey was conducted in February 2013 as an e-enquiry. The respondents included 166 entrepreneurs with full-time or part-time businesses. The respondents provided riding services (36% of respondents), livery stables (28%), trotter training (14.5%), breeding (11.5%) and tourism and welfare services (10%). Almost half of the respondents (48%) had been in business for 6-20 years. In addition, 17% of the respondents had been in business for more than 20 years and 35% for less than five years. There were approximately 19 horses per stable.

The central topics of the survey were: (1) the equine business in Finland (whether it is growing or not); (2) demand; and (3) profitability (how the entrepreneurs see it now and in the near future, i.e. within the next five years). The assessment used to evaluate demand and profitability in the equine industry was six-scaled, in descending order: excellent, good, fair, acceptable, rather poor or poor. The scale for estimating profitability was the same. The estimates were based on how the entrepreneurs reported feeling about the situation; they were not given any numbers or facts to evaluate profitability.

There was also a question about how entrepreneurs estimate the present state of equine industry. The assumption behind the question was that the estimates of the present state affect future plans and entrepreneurs' strategic choices. The subjects were given eight topics to consider: environmental issues in stables, horse's welfare and horsemanship, interest in horse ownership, the image of and communication in the horse sector, amount and quality of domestic breeding, entrepreneurs' business skills, situation of youth work and welfare of entrepreneurs and respect towards their work. The scale used to evaluate the present state of each issue had the values of excellent (grade 6), good (grade 5), fair (4), acceptable (3), rather poor (2) or poor (1). No exact definitions were given for the issues in question, but it is likely that the respondents understood them equally well.

In addition to the survey, workshops were held with 20 participants, entrepreneurs or other professionals in the equine sector. The workshops included structured discussions about the Finnish equine industry and its future. The topics were the same as they were in the survey: is the equine business in Finland growing or not and do entrepreneurs see an increase in demand and profitability in the near future, i.e. within the next five years. In addition to these topics there were discussions on the marketing of the horse sector. These subjects led to discussions about the strengths, weaknesses, possibilities and threats of the horse sector but also discussions about the values of and motives to work in the horse business.

Results

Almost half of the respondents (45%) stated that they will enlarge their businesses within the next five years (Figure 1). Those who reported having plans to enlarge their business were typically younger entrepreneurs with new enterprises: 48% of the respondents had carried out their business for less than five years and the average age of the entrepreneurs was 39 years. The reasons for enlarging their

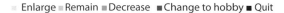

Enlarge ■ Remain ■ Decrease ■ Change to hobby ■ Quit

All (166)	45%	33%	13%	5%	4%
Riding services (60)	43%	32%	13%	5%	7%
Livery stables (46)	54%	35%	4%	4%	2%
Trotter training (24)	17%	38%	33%	4%	8%
Breeding (19)	47%	32%	11%		11%
Tourism/welfare sector (17)	65%	24%	12%		

Figure 1. Entrepreneurs' (n=166) plans to enlarge or reduce their businesses until the year 2018. Almost half of all respondents had plans to enlarge their businesses.

businesses include strategic planning, increasing demand or generational changes in the business (e.g. an aged owner retiring from the business). The number of respondents who stated that they have plans to decrease or to quit business was fewer than 2% per year and has not changed significantly compared to the findings from an earlier study (Pussinen and Thuneberg, 2010). In most cases, the reasons for quitting had to do with lack of profitability, retirement, lack of money to invest in the business or changes in the owner's health. The average age of those who were quitting was 55 years.

Generally, most of the entrepreneurs (70%) feel that the demand will be excellent or good in the next five years (Figure 2). Only the trotter training and breeding sectors were not as positive as the other sectors. The reasons for pessimism in these sectors may have to do with the state of the world economy and also the age of the enterprises. Entrepreneurs in the trotting and breeding sectors are typically older than in other sectors: one third of the respondents (35%) had been entrepreneurs in the breeding or trotting sectors for a long time, i.e. more than 20 years.

Nearly half of all the entrepreneurs (43%) expect that profitability will either be excellent or good in the next five years (Figure 3). The entrepreneurs who manage riding stables and work in the tourism and welfare sector expect that profitability will be rather good, but the trotter training, livery stables and breeding entrepreneurs reported feeling more uncertain about the future. More than one tenth of respondents (14%) estimate the profitability will be rather poor or poor. the entrepreneurs in tourism or the welfare sector have a growing interest and faith in good profitability, but it is notable that the businesses are mostly small-scale and young at present.

One question in the given topics dealt with the present state of the horse sector (Figure 4). Due to the fact that the present state affects the future plans of the horse business and gives ideas to further development of the equine industry, results are needed for further education and development projects. The entrepreneurs estimated environmental issues, for example the state of fences and buildings, and the horse's welfare to be quite good (the average grade was fair 4.0). The worst average grade (3.0) was given to the entrepreneur's own welfare and feelings about respect towards the work. The situation in youth work, meaning informal activity and clubs for children, was given the grade of 3.1. The amount and quality of horse breeding was estimated acceptable, as was the entrepreneur's business skills. Interest in horse ownership and the image of the horse sector were seen as quite fair.

After the survey, we discussed the results in workshops with entrepreneurs or other professionals in the equine sector. The survey showed there is a worry about the image of the horse sector. That is why one of the workshop conversations addressed the question 'What should we be proud of in the horse sector'? The answers mostly dealt with horses, educated people and horses' effect on human

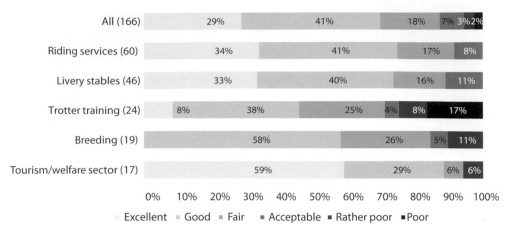

Figure 2. Entrepreneurs' (n=166) estimates of demand until 2018. Most (70%) of the respondents estimated that demand will develop at an excellent or good level.

welfare. Free-form answers for the question included: 'We should be proud of horses and knowledge of the professionals. Taking care of horses teaches responsibility and empathy to youngster.' 'A

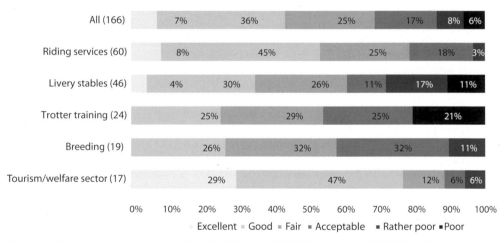

Figure 3. Entrepreneurs' estimates of profitability until 2018. Less than half (43%) of the respondents estimated that profitability will develop at an excellent or good level.

versatile national horse breed, the Finnish horse, connects the riding and trotting sectors. Horses are beautiful animals, feelings of success with them and a sense of solidarity keeps on going for a long time.' 'Horses are therapeutic animals, making us exercise outdoors and find new friends', 'Animals, especially horses are windows of the soul for people. You can trust that someone who gets well along with horses cannot be a bad person', 'We can be proud of professional skills and of Finnish traditional and at the same time modern education system.' These answers show that there is considerable knowledge in the horse sector but the problem is how to transfer this knowledge to the wider audience.

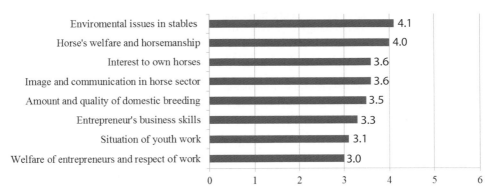

Figure 4. Entrepreneurs' (n=166) estimates of the state in 2013 of the equine industry in Finland. The numerical values give the average grade for each topic. The scale used to evaluate the present state consist of the values of excellent (grade 6), good (grade 5), fair (4), acceptable (3), rather poor (2) or poor (1).

The other main task was to analyse the horse sector using a SWOT-analysis. The participants in the workshops emphasised that the main strength of the sector is the level of enthusiasm of the horse owners and entrepreneurs. The tendency of treating horse business as a lifestyle can be seen as a strong motivator, but it also brings challenges to profitability. Such challenges are business weaknesses, which highlights the fact that horse professionals need more business skills and know-how about pricing, marketing and financial planning. If business is not profitable, it is not possible to invest in or take care of the entrepreneur's wellbeing. Marketing and brand management also give opportunities to small-scale businesses. Threats to the horse business include the state of the world economy, because that affects the amount of money that customers can spend. Also, the media can greatly affect the attitudes of the general public. The effect may be positive or negative. The well-being of horses and safety of the environment and facilities may attract publicity. The sport and racing sectors, especially the superstars, interest the public.

Conclusions

As a conclusion, we can say that services and entrepreneurship in the equine industry are undergoing change. When analysing the results from a wider aspect, we can see that there are new challenges related to the entrepreneur's competence. Horses interest people and the equine industry in Finland still seems to be growing. Growing demand, however, is not a self-evident truth. There are many great things in the horse sector but the question remains, how to convince people who are unaware of them? Feelings, stories, personalities of people and horses are the best base for improving marketing communication. These are the ways to reach new customers for the world of horses. Customers of the future might like other things than those we in the horse sector are used to. For some segments, for example for older people, just looking at horses or taking care of them might be enjoyable.

Results show that horses themselves are the main motivating factor for professionals. It is interesting that people inside horse world can see the many positive things there. But do outsiders see all the positive aspects in the same way? The image of the whole horse sector is affected by how the environment and horses are perceived, how clean the surroundings are and how the customers are taken care of. Things to be proud of that we came across in our workshops discussion should be used strongly in marketing. The aim of commercial marketing is to create demand and profitability. In addition to commercial marketing there is a possibility to increase the attraction of the horse sector by using methods of social marketing. Social marketing seeks to develop and integrate marketing concepts with other approaches to influence behaviours that benefit individuals and communities for the greater social good (Kotler and Armstrong, 2014). In the horse sector, social marketing could be based horses' impacts on society and human welfare. The horse sector employs thousands of professionals and is a significant part of the rural landscape.

According to our workshop conversations, new customers, especially adults, demand more quality. They value the way how they are treated. For entrepreneurs, it is not enough to take care of horses, social skills are also needed. This aspect led us to assess the skills the future entrepreneur needs. It is well known that the motives of someone to engage in horse business are not only economic. Horse business is seen as a way of life now, but in the future this might be different. To succeed, entrepreneurs will need to have more economic aims and be more competitive. On the other hand, they do not need to do everything by themselves. It is more important to know where help and consulting can be found, for example for marketing and accounting. Networks and co-operation with other entrepreneurs may help in this. The values of the business and its quality-based aims typically include being in a dream job or living in a rural area. These values cannot be sacrificed for the sake of economical aims. Qualitative motives describe the type of entrepreneur in horse business and make him/her 'unique'. Money is not everything.

There are many opportunities for new services in new equine industry. Entrepreneurs need courage to overhaul their business and analyse the needs of new customers. Different kinds of customer segments, including children, young people or seniors, all need different kinds of services. New customer groups may press for specialised services and challenge the entrepreneur's business skills. For entrepreneurs, there is a need to retain and create more demand, given the competition between multiple pastime activities and the fact that people do not always have plenty of leisure time. Like any branch of business, the equine industry needs more marketing and brand building to find new customers. Basic business strategies, such as differentiation or focus strategies, could be adapted to the equine industry, just as is the case in any other business. With the differentiation strategy, businesses concentrate on achieving superior performance in an important customer benefit area valued by a large part of the market. With the focus strategy, a business focuses on one or more narrow market segments (Kotler and Armstrong, 2014).

According to the survey, demand and profitability in equine enterprises seem to correlate: however the entrepreneurs estimated that demand will develop better than profitability. This leads to the question: How can they improve profitability? The process leads to a fair pricing policy. Pricing can be seen as one of the most important components of profitability. When a business is well run, it is possible to invest in facilities, horses and staff. And that improves the quality of the business, leaving customers satisfied and happy. Pricing is not easy in the equine industry. All businesses operate within their own individual set of circumstances and therefore have unique cost and income structures (Eastwood et al., 2006). The size of the stable will likely have an impact on the fixed costs per horse. It is not enough to define the price like others do in the same area. Also, the relationship between the customer and entrepreneur can be personal, almost like a friendship. Prices should be increased when costs increase. This is not done on a regular basis. Excessively low prices lead to poor profitability.

So how can equine industry professionals adapt to new customer demands and be successful in the future? Equine entrepreneurs should be encouraged to think on a big enough scale and to see new possibilities, to know about specialised services which they can offer, and be able to identify new, not yet known, potential customers. There is a need for further education for entrepreneurs and those who have plans to start businesses in this area. Life cycles of the enterprises are different: there are entrepreneurs who have been in the horse sector for a long time. They need further education and consulting for example, in product development and knowledge about how to hire employees. Those who are starting their business need more skills related to the establishment of the business and to profitability analysis. Business skills strengthen the whole equine industry's competitiveness and help entrepreneurs provide high-quality services. Co-operation between different kinds of enterprises, such as riding services, trotter training, livery stables, tourism and the welfare sector is almost non-existent at the moment. The aim of co-operation should be to maintain existing customers in the 'horse world'. Like a 'lifelong hobby' horses provide leisure-time activities for all customers.

The basis for further research and development projects is to see the big picture: it is important to see the whole equine industry internationally, nationally and regionally. On the whole, there is a need to develop co-operation between research, consulting and education. There are growing needs to do more research on the equine business's economy and customer-based markets.

References

Eastwood, S., A. Riis Jensen and A. Jordon, 2006. Business management for the equine industry. Blackwell Publishing Ltd, Oxford, United Kingdom, 158 pp.

Hippolis ry, Suomen Hippos ry, Suomen Ratsastajainliitto ry and MTT Hevostalous, 2014. Hevostalous lukuina 2013. Available at: www.hippos.fi/files/8773/Hevostalous_lukuina_2013_web.pdf.

Hippolis ry, 2010. Suomalaisen Hevosalan katsaus 2010. Available at: http://hippolis.fi/UserFiles/hippolis/File/Hevosalankatsaus_uusin_pienempi.pdf.

Kotler, P. and G. Armstrong, 2014. Principles of marketing. Pearson Education Limited, New York, NY, USA, 716 pp.

Liljenstolpe, C. 2009. Horses in Europe. Swedish University of Agricultural Sciences (SLU), Uppsala, Sweden. Available at: http://tinyurl.com/qa8ghj9.

Pussinen, S. and T. Thuneberg, 2010. Katsaus hevosalan yritystoimintaan raportti hevosyrittäjyys 2009 – kyselystä. HAMK Publications, Tampere, Finland, 48 pp.

Part 5.

Social economy of the equine sector

Part 5 looks at three interesting subsectors of the equine economy which might be considered to be innovative – social media, complementary medicine for horses, and equine tourism. All provide a detailed look at how equine entrepreneurs (and riders in general) are bringing innovations in from the wider society and employing them for their own economic purposes. The first looks at the motivations of tourists who wish to ride horses. Understanding the way in which people's passion for horses works can underpin innovative ways of marketing equine tourism services, and devising new types of tourism product itself. Just like in the society which surrounds the sector, horse people use social media more and more. In a very interesting study, another author compares social media use by those who run agricultural enterprises and those who run equine businesses. She finds some distinctions, some of which are built around professionalism, and the specialized nature of equine knowledge. Another author reports on the growing sector of 'alternative medicine' being increasingly supplied to the equine sector, for both rider, and horse. She concludes that there are many opportunities within the subsector, but that there are problems with regulation and credibility.

10. Does age and horse ownership affect riders' motivation?

J. Wu*, G. Saxena and C. Jayawardhena
The University of Hull, Business School, Cottingham Road, Hull, HU6 7RX, United Kingdom; jie. wy@163.com

Abstract

This chapter examines underlying reasons for leisure riders' participation in riding activities. It draws on Iso-Ahola's (1982) motivation theory to understand key tenets like escape, learning, relaxation, social and self-esteem underpinning leisure riding behaviour. We examine not only the leisure riding motivation, but also whether there are any differences in the motivations between horse owners and non-horse owners, and extend the inquiry among riders of different age groups. Data were collected utilising an online survey comprising of 186 respondents. Findings suggest that riders' age has a significant influence on their motivation to undertake leisure riding while horse ownership does not appear to have a significant impact. Also, results highlight the key role of relaxation and learning underlying leisure riding.

Keywords: riders, motivation, age

Introduction

Riding activities have enormous contribution to social, cultural and national economies around the world. Riding attracts both horse owners and significant numbers of non-horse owners, ranging from beginner to advanced, covering any age of person from under five to retired(Jodd, 1994). However, this vital segment remains under researched in terms of their motivations. Motivations in leisure context have been referred to as psychological or biological needs and wants (Crompton and McKay, 1997).There are few published analysis (Daniels and Norman, 2005) and descriptions of riders' motivations (Cynarski and Obodyński, 2008; Buchmann, 2014; Helgadottir, 2006). Daniels and Norman's study (2005) on the motivations of spectators in a racing event and related research does not consider or examine the riders' motivation. Duff-Riddell and Louw(2011) investigate young female competitive horse riders and does not analyse leisure riders' motivations. Ellison (2009) explores how eight contemporary horsewomen perform femininity in their daily lives using ethnographic research methods. Buchmann (2014) exploratory research describes riders' motivation as 'being able to take time off work, getting away from it all, to have the opportunity to progress in horsemanship, to meet like-minded people and network' (P11).A review of literature indicates a lack of quantitative study on motivations of leisure riders. Thus the focus of this study in quantitative analysing the motivations of riders as core consumers is likely to enhance an understanding of leisure riding as a product.

Motivations of leisure riders

Leisure motivation is commonly seen as the driving force behind all leisure activities(Crompton 1979, Iso-Ahola and Allen 1982, Snepenger *et al.* 2006).A generally accepted definition of motivation comes from Murray(1964, p. 7), who stated that '[a] motive is an internal factor that arouses, directs, and integrates a person's behaviour.' Motivations in leisure context have been referred to as psychological or biological needs and wants (Crompton and McKay, 1997). Thus a better understanding of leisure motivations is critical for marketers in understanding the leisure patterns of consumers (Crompton, 1979; Snepenger *et al.*, 2006).

Iso-Ahola's (1982) escape-seeking theory categorised leisure motivation forces into two main factors, namely, obtaining psychological rewards through travel in a different destination and escape associated with the desire to get away from routine environments. An individual engages in a leisure activity to gain intrinsic rewards (such as friendship, love) and seek an escape from routine environments or urban stress (Iso-Ahola, 1983). Both escape and seeking motivation influence individuals' participation in a leisure activity simultaneously (Iso-Ahola, 1983, 1990). Iso-Ahola (1990) further suggests that both dimensions have personal (psychological) and interpersonal (social) components. Personal motivations include rest/relaxation, prestige, competence, learning about other cultures while interpersonal motives include social interaction, encounter with natives, strange cultures and past life styles. This research draws on Iso-Ahola's motivation theory with reference to recent relevant research to investigate riders' motivations.

Broadly, four motivation factors that have been extensively used in leisure and tourism motivation research include escape, social bonding, learning/relaxation and self-development (Pearce and Lee, 2005). Within leisure riding context, those travel motivation factors are supposed as leisure riders' motivations to participate in riding related activities.

Learning and relaxation could be important when riders participate in riding activities. Riders may take time to pursue activities of interest and learning something new. People could learn life lessons from horses, such as non-confrontation; planning ahead; patience; persistence; consistency; fix it and move on (Rashid 2011).In addition, self-development or self-esteem could be another essential motivation factors for riders to pursue riding activities (Cynarski and Obodyński, 2008). Klish (2009) investigates of adolescent girls' experiences of riding horses focuses on four relationships with self, the barn community, horse and trainer that connection to adolescent girls' development. Horses provide an authentic mirror to the self, reflecting qualities of the rider that she or he may or may not wish to view. Owners learn the subtle art of persuasion and an awareness of how and when to bestow rewards or remove pressure – skills that they may transfer in their interface with their peers and friends (Keaveney, 2008).

Riding could provide riders an opportunity to escape from the daily routine. Horse riding is a popular recreational activity throughout the world. In many countries, especially developed ones, protected areas such as national and regional parks and wilderness areas, are recognised and used for recreational horse riding (Newsome et al. 2008). Further, while analysing motives for horse riding, attention could be paid to realizing social bonding and relationship enhancement (Cynarski and Obodyński 2008). Many young people like to work with and ride horses, hence the widespread occurrence of pony clubs, equestrian centres, riding schools and horse riding magazines. These places give people a chance to meet new people in different locations or to be with people who have similar interests.

Riders' motivations difference

Socio-demographic variables are the primary factors that influence people's leisure-related behaviour (Jang and Wu, 2006; Zimmer et al., 1995). These variables may include age, gender (Jönsson and Devonish, 2008), education, economic and health status and nationality/cultural background (Kim and Prideaux, 2005; Maoz, 2007). The literature review indicates that there may be a combination of socio-demographic variables that influence leisure and tourism behaviours. The results of previous studies imply that different sample groups may have different socio-demographic factors influencing their behaviours. As little is known regarding leisure riding behaviour of riders, understanding the demographic variables affecting their leisure riding behaviour should help leisure riding enterprises to develop effective marketing programmes that could better satisfy the leisure needs and expectations of their consumers.

Gender differences in riding activities has been emphasised by numerous research (Calamatta, 2012; Dashper, 2012). Males are successful at the higher status level of competition and females particularly participate in the lower status positions as grooms (Calamatta, 2012). Social role theory is frequently employed in explaining gender-related differences in many fields (Saad and Gill, 2001). According to social role theory, women and men play different roles and exhibit dissimilar behaviours in society because they are differently socialized (Saad and Gill, 2001). In fact, women dominate the horse industry, especially for leisure riding and in riding schools (Chevalier and Grefe, 2013). Calamatta's (2012) study on gender and work in a British riding school explains clearly that woman in their nurturing role as opposed to men who are able to maintain an emotional distance from the horses, are predominant consumers and providers of riding related activities. This research confirms Calamatta's research on leisure riding, dominated by females. Gender difference is not the primary concern in this research.

Age is a key demographic category in leisure riding context. Age has been found to have a curvilinear relationship with leisure time (Thrane, 2000). Age differences in leisure tourism preferences are most readily explained with reference to life course expectations (Gibson and Yiannakis, 2002). The majority of riders in riding school are young and still live at home with their parents (Calamatta, 2012). It is important to differentiate between those individuals who are young (less than 20 years old) and those who would be classified as mature (age 20 and older) to see if differences existed in terms of leisure riding motivations.

Horse ownership is a special segmentation variable in riding context. Horse owners may have more disposable income or have a high social class. Higher levels of income are deemed to enable higher levels of leisure and tourism consumption (Greenwell *et al.* 2002). Higher income renders greater discretionary spending capability, with the result that it is easier to purchase and board horse(s). Consequently, amongst horse owners, the threshold required to participate in riding activities should decrease. Therefore, this study compares leisure riders' motivations according to horse ownership and age of respondents.

Methodology

This study utilised survey methodology and online data collection. Data was collected from sample units in the county of Yorkshire, UK. They included two equestrian centres providing riding lessons and riding camps, two higher education institutions having equine subject and one leisure riding agency that organises trail riding within Yorkshire. Riders' motivations were measured through 14 items taken from previous tourism (Hung and Petrick, 2011) and leisure research (Kyle *et al.*, 2006) to ensure consistency with prior studies. As is the usual practice, measures were adapted to the leisure riding research context. This research uses five points Likert scale as a measure technique (where 1=strongly agree and 5=strongly disagree).

The questionnaire was pretested through face to face interviews with riders at an equestrian centre in Sep, 2012. Revised questionnaire was uploaded on survey monkey and advertised by Facebook webpages of each sample units. The main online survey was conducted from Nov, 2012 to Feb, 2013. Out of a total of 211 respondents of 186 (88.2%) were considered usable and 25 were discarded due to an incompleteness of responses. The sample size and percentage of each subgroup are illustrated in Table 1.

Results

Descriptive statistics were conducted to determine the mean and standard deviation scores for each motivation items. Almost all items obtained mean scores below the mid-scale point of 3, which indicated that majority of the measured items, corresponded with riders' motivations. The only exception was one motivation item 'To do something that impresses others' with the highest mean value of 3.52, indicated that generally respondents do not participate in riding to impress others.

Table 1. Demographic profile of respondents.

		Frequency (n=)	Percentage
Total		186	100.0%
Own horse(s)	No	63	33.9%
	Yes	123	66.1%
Age	Less than 20 years old	87	46.8%
	20-30 years old	30	16.1%
	31-40 years old	29	15.6%
	Over 40 years old	40	21.5%
Respondents	Riding schools or equestrian centres	68	36.6%
	Students training in equine science	95	51.1%
	Tourists engaged in trail riding/farm stays	23	12.4%

Difference of riders' motivations with different demographic backgrounds were compared using Independent sample T-tests and one-way ANOVA. Respondents were separated into several groups according to different age and horse(s) ownership first, and then the mean scores of the different constructs for each group were compared.

Analysis of riders motivations by age

An ANOVA was conducted to identity whether riders' motivations were significantly different within different age groups. Respondents were separated into four groups (less than 20 years old, 20-30 years old, 31-40 years old and over 40 years old).The result of ANOVA tests (Table 2) indicated that there were significant differences in two escape items (sig=0.00 and 0.01), one learning item (sig=0.01), three items of relaxation (sig=0.01, 0.01 and 0) and two social items (sig=0.02) among riders with different age.

- Group 1. Less than 20 years old: the results of the ANOVA showed that a statistically significant difference (sig=0.01) exists between riders less than 20 years old (group 1) and older than 40 years old (group 4) on the learning item 'To develop my knowledge about riding'. Group one (mean=1.48) had greater interest in learning riding knowledge than group 4 (mean=1.9). Additionally, group one was more motivated by relaxation and social than other groups. Group one agreed more with the statement that their participation in riding related activities was due to their desire 'To do exciting things' (mean=1.61) than the group 4 (mean=2.00), 'To have fun, be entertained' (mean=1.44) than group 2 (mean=1.87), and 'To enjoy riding that provides a thrill' (mean=1.44) than both group 3 (mean=1.83) and group 4 (mean=2.05). Similarly, group one agreed more with the statement -'To be with people who have similar interests' (mean=1.78) than group 2 and 4 (mean=2.33 and 2.18), and -'To do something with my family/friends' (mean=2.26) than group2 (mean=2.26).
- Group 2. 20 to 30 years old: the results of the ANOVA showed that a statistically significant difference exists between riders 20 to 30 years old (group 2) and older than 40 years old (group 4) on three items. Group two were more motivated by relaxation 'To enjoy riding that provides a thrill' (mean=1.63) than group 4 (mean=2.05).
- Group 3. 31 to 40 years old: the results of the ANOVA showed that a statistically significant difference exists between riders 31 to 40 years old (group 3) and less than 20 years old (group 1) on escape motivations. Group 3 were more motivated by escape 'To be away from the everyday

Table 2. One-way ANOVAs on riders motivations by age.[1,2]

Motivation	Item	Group 1	Group 2	Group 3	Group 4	F	P-value[3]	Difference[4]
Escape	1. To be away from the everyday routine of home	2.53	2.43	1.90	1.95	4.961	<0.001*	1>3,4
	2. To get away from crowded areas	2.67	2.70	2.14	2.23	4.220	0.01*	1>3,4
	3. To experience the solitude/privacy of riding	2.18	2.63	2.24	2.20	2.036	0.110	N/A
Learning	4. To develop my knowledge about riding	1.48	1.63	1.86	1.90	3.795	0.01*	4>1
	5. To learn more about horsemanship	1.53	1.63	1.72	1.75	1.272	0.285	N/A
Relaxation	6. To do exciting things	1.61	1.87	1.93	2.00	3.691	0.01*	4>1
	7. To have fun, be entertained	1.44	1.87	1.41	1.58	3.917	0.01*	2>1,3
	8. To enjoy riding that provides a thrill	1.44	1.63	1.83	2.05	8.491	<0.001*	4,3>1; 4>2
Social	9. To be with people who have similar interests	1.78	2.33	2.10	2.18	5.159	<0.001*	4,2>1
	10. To talk to new and varied people	2.24	2.43	2.66	2.53	2.247	0.084	N/A
	11. To do something with my family/ friends	2.26	2.87	2.59	2.53	3.349	0.02*	2>1
Self-development	12. To do something that impresses others	3.21	3.73	3.69	3.93	5.488	<0.001*	4>1
	13. To help me feel like a better person	2.52	2.43	2.72	2.73	0.781	0.506	N/A
	14. To derive a feeling of accomplishment	1.92	1.80	1.97	1.93	0.300	0.826	N/A

[1] Group 1: less than 20 years old; Group 2: 20-30 years old; Group 3: 31-40 years old; Group 4: more than 40 years old.
[2] Items measured along a 5-point scale where 1=strongly agree and 5=strongly disagree.
[3] * = $P<0.05$.
[4] n/a = not available

routine of home' (mean=1.9) and 'To get away from crowded areas' (mean=2.14) than group 1 (mean=2.53 and 2.67). Same with group 1, group 3 were more concern about 'To have fun, be entertained' (mean=1.41) than group 2 (mean=1.87).

- Group 4. Over 40 years old: the results of the ANOVA showed that a statistically significant difference exists between riders over 40 years old (group 4) and less than 20 years old (group 1) on escape motivations. Similarly with group 3, group 4 were more motivated by escape 'To be away from the everyday routine of home' (mean=1.95) and 'To get away from crowded areas' (mean=2.33) than group 1 (mean=2.53 and 2.67). Further, group 4 were more disagree to 'To do something that impresses others' (mean=3.93) than group 1 (mean=3.21).

In summary, the data result show that younger groups (group 1 and 2) were more motivated by learning, relaxation, social while the mature groups were more motivated by escape.

Analysis of riders motivations by horse ownership

An independent T-test was conducted to identify whether horse owners' motivations were significant different with motivations of those people do not own a horse. The respondents were separated into horse owner group (n=123) and do not own horse(s) group (n=63). The mean scores of motivations for each group were compared by T-test statistics (Table 3). The result indicated that significant difference exists between two groups on two motivation items – 'To enjoy riding that provides a thrill' (P=0.03), 'To do something with my family/friends' (P=0.01).

The group who do not own a horse agreed more with 'To enjoy riding that provides a thrill' (mean=1.65) and 'To do something with my family/friends' (mean=2.38) than the other horse owner group (mean=1.667 and 2.512, respectively). Therefore, those who do not own a horse were more likely to be motivated by doing something with their family/friends, feel thrill when riding than horse owner group. For the other items, no significant differences were found. This suggests that apart from two items ('To enjoy riding that provides a thrill' and 'To do something with my family/

Table 3. Independent t-tests on riders motivations by horse(s) ownership.[1]

Motivation	Item	Do not own horse (n=63)		Horse owner (n=123)		Mean diff.	T-value	P-value[2]
		Mean	SD	Mean	SD			
Escape	1. To be away from the everyday routine of home	2.44	1.118	2.211	0.969	0.233	1.472	0.106
	2. To get away from crowded areas	2.56	0.963	2.463	0.917	0.092	0.637	0.933
	3. To experience the solitude/ privacy of riding	2.24	0.837	2.285	0.928	-0.046	-0.334	0.803
Learning	4. To develop my knowledge about riding	1.67	0.741	1.650	0.768	0.016	0.138	0.422
	5. To learn more about horsemanship	1.63	0.630	1.618	0.696	0.017	0.163	0.310
Relaxation	6. To do exciting things	1.75	0.718	1.805	0.709	-0.059	-0.534	0.797
	7. To have fun, be entertained	1.60	0.661	1.496	0.632	0.107	1.078	0.827
	8. To enjoy riding that provides a thrill	1.65	0.626	1.667	0.743	-0.016	-0.145	0.03*
Social	9. To be with people who have similar interests	1.95	0.851	2.033	0.757	-0.080	-0.655	0.937
	10. To talk to new and varied people	2.38	0.869	2.407	0.838	-0.026	-0.194	0.816
	11. To do something with my family/friends	2.38	0.812	2.512	1.027	-0.131	-0.882	0.01*
Self-development	12. To do something that impresses others	3.51	1.014	3.528	1.089	-0.021	-0.124	0.439
	13. To help me feel like a better person	2.60	1.040	2.569	1.009	0.034	0.216	0.693
	14. To derive a feeling of accomplishment	2.14	0.737	1.789	0.681	0.354	3.265	0.611

[1] Items measured along a 5-point scale where 1=strongly agree and 5=strongly disagree. SD: standard deviation.
[2] * = P<0.05.

friends') both rider groups [i.e. (non) horse owners]of this study had similar motivations to engage in riding related activities.

Results from principal components analysis

This study employed principal components analysis (PCA) to extract meaningful motivation factors since PCA is most appropriate when data reduction is a primary concern (Field, 2009). Further, factor rotated by Quartimax method. Because of using Quartimax rotation, a variable loads high on one factor and as low as possible on all other factors so that interpreting variables becomes easier (Hair et al., 2010).Three criterions for factors extraction include: (1) factors have eigenvalues greater than 1 (Kaiser, 1970); (2) items have factor loading for one factor greater than 0.45 (Hair et al., 2010); and (3) the reliability of factors (Cronbach's Alpha) are greater than 0.7 (Field, 2009).

The final PCA was conducted on five items (Table 4). The estimates of KMO statistics of 0.697, is above the acceptable limit of 0.5 (Field, 2009). Bartlett's test of sphericity $\chi^2(10)=234.552$, $P<0.001$, indicated that PCA was appropriate for those motivation items. Two components had eignvalues over Kaiser's criterion of 1 and in combination explained 70.51% of the variance. Given the convergence of the screen plot and Kaiser's criterion on two components, this is the number of components that were retained in the final analysis. Table 4 shows the factor loading after rotation. The items that cluster on the same components suggest that factor 1 represents relaxation and factor 2 learning.

Factor one, relaxation, explains 39.98% of the variance with an eigenvalue of 2.00. The relatively large proportion of the total variance for this factor leads us to conclude that among leisure riders, relaxation represents a central distinguishing motivational theme. The reliability alpha of this factor is 0.72, indicated an accepted reliability.

Factor two, learning, explains 30.54% of the variance with an eigenvalue of 1.53. The reliability alpha of this factor is 0.74, indicated an accepted reliability. Further, it was confirmed by the Pearson correlation of 0.59, which is statistically positive and significant at 0.01 level, indicating that the factor is reasonably reliable even with only two items kept.

Table 4. Motivation factors of leisure riders.[1]

	Component	
	1	2
To do exciting things	0.824	
To have fun, be entertained	0.771	
To enjoy riding that provides a thrill	0.769	
To develop my knowledge about riding		0.901
To learn more about horsemanship		0.814
Eigen value	2.00	1.53
Variance explained (%)	39.98	30.54
Cronbach's alpha	0.72	0.74

[1] Kaiser-Meyer-Olkin measure of sampling adequacy = 0.697. Chi-square(10) = 234.552. Bartlett's test of sphericity, $P<0.001$. Total variance explained 70.51%.

Discussion

This study has revealed that there are significant differences based on age difference regarding leisure riders' motivations. Younger riders (less than 30 years old) are motivated more by learning, relaxation, social while the mature riders (over 30 years old) are motivated more by escape. In the tourism context, escape means a temporary change of environment, not only from the general residential locale but also from specific home and job environments (Prebensen et al., 2010). Crompton (1979) explains escapism as a key motivation underlying individuals' need to relax include the need for peace and quiet. Horse riding in outdoor settings, particularly those perceived as natural which generally linked with a peace and quiet environment. These environments enable riders to escape from daily routine and seek comfort at the same time. Consequently, mature riders may require a higher threshold of motivation than younger riders to attend a riding activity.

With respect to horse ownership difference, the results of the study revealed that the two groups of leisure riders' (do not own a horse and horse owners) did not differ significantly in their motivations to leisure riding. This finding is different with horse tourists' behaviour research. Tourists hiring horses are always wanted to ride and tourists who own horses do travel short (Buchmann, 2014).

Findings highlight relaxation and learning as the most important motivations for riders participated in leisure riding-related activities. Relaxation motivation of leisure riding is consistent with other research on leisure activities (e.g. spa breaks (Mak et al., 2009), theme-parks based entertainment (Park et al., 2009)). Similar to majority of other leisure activities, riders spend time to pursue activities that are of intense personal interest and help them to relax. Relaxation deals with do things that make an individual escape from civilization and from routine and responsibility (Crandall, 1980). Moreover, the learning motivation is relatively less exactly mentioned in other leisure or tourism motivation research, like, sport tourism (Funk and Bruun, 2007), sun and sand tourist vacation (Prebensen et al., 2010). Helgadottir (2006) points out that the urge to learn horsemanship culture is a unique motivation for riders to engage in riding related activities. People can learn life lessons from horses, like communication, self-confidence and self-image, concern and compassion empathy (Rashid, 2011; Witter, 2000).

A limitation of this study has been the dependency on an online survey which means that the sample does not necessarily represent the leisure rider population accurately. Also, online survey eliminated the participation of those with no internet access and computer skills. However, a key advantage of the online survey is that it encourages honest feedback; lower cost of research; no interviewer bias; desirability; high speed, instantaneous results of polling (Matthews and Ross, 2010). Another limitation is the choice of leisure riding motivation factors and of the variables that represent these motivation factors. Selection biases could also affect the results as the sample includes a lot of young participants riding in riding schools and the sample is not representative of the whole riders' population.

Conclusion

To conclude, it can be said that this study has revealed previously under explored leisure riders' motivations using a quantitative study. The study concludes that age should be considered in predicting variation in leisure riders' motivations while riders with different horse ownership do not have significant impact on their leisure riding motivations. Findings suggest that relaxation and learning are two main motivations for riders participating in riding activities. A better understanding of motivations in rider pursuits may help leisure riding enterprises managers to build loyalty and repeat business in a destination. Since the respondents of this survey are mainly British, future studies could compare respondents from different countries to examine nationally effect on riders' motivations. Furthermore, research could be conducted on what destination attributes leisure seekers chose, their preferred choice, and what images they have of leisure riding destinations. This would help destination marketers understand how they can position themselves and develop more effective marketing communication activities.

Reference

Buchmann, A., 2014. Insights into domestic horse tourism. The case study of Lake Macquarie, NSW, Australia. Current Issues in Tourism http://dx.doi.org/10.1080/13683500.2014.887058.

Calamatta, K., 2012. A Qualitative study of gender and work in a British riding school. Doctor of Philosophy, University of Sussex, Brighton, United Kingdom, 291 pp.

Chevalier, S. and G. Grefe, 2013. Equine business. the spectacular growth of a new equine economy in France. In: Book of abstracts of the 64th Annual Meeting of European Federation of Animal Science. Wageningen Academic Publishers, Wageningen, the Netherlands, p. 248.

Crandall, R., 1980. Motivations for leisure. Journal of Leisure Research 12(1): 45-54.

Crompton, J.L., 1979. Motivations for pleasure vacation. Annals of Tourism Research 6(4): 408-424.

Crompton, J.L. and S.L. McKay, 1997. Motives of visitors attending festival events. Annals of Tourism Research 24(2): 425-439.

Cynarski, W.J. and K. Obodyński, 2008. Horse-riding in the physical education, recreation and tourism – axiological reflection. Research Yearbook 14(1): 37-43.

Dashper, K.L., 2012. Together, yet still not equal? Sex integration in equestrian sport. Asia-Pacific Journal of Health, Sport and Physical Education 3(3): 213-225.

Daniels, M. and W. Norman, 2005. Motivations of equestrian tourists. An analysis of the colonial cup races. Journal of Sport & Tourism 10(3): 201-210.

Duff-Riddell, C. and J. Louw, 2011. Achievement goal profiles, trait-anxiety and state-emotion of young female competitive horse riders. South African Journal for Research in Sport, Physical Education & Recreation 33(3): 37-49.

Ellison, S., 2009. Towards the horsewoman. Performing femininity in the American horse training and riding arenas. Bowling Green State University, Bowling Green, OH, USA, 253 pp.

Field, A., 2009. Discovering statistics using SPSS for Windows advanced techniques for the beginner. SAGE, London, United Kingdom, 496 pp.

Funk, D.C. and T.J. Bruun, 2007. The role of socio-psychological and culture-education motives in marketing international sport tourism. A cross-cultural perspective. Tourism Management 28(3): 806-819.

Gibson, H. and A. Yiannakis, 2002. Tourist roles. Needs and the lifecourse. Annals of Tourism Research 29: 358-383.

Greenwell, T.C., J.S. Fink and D.L. Pastore, 2002. Perceptions of the service experience using demographic and psychographic variables to identify customer segments. Sport Marketing Quarterly 11: 233-241.

Hair, J.F., W.C. Black, B.J. Babin and R.E. Anderson, 2010. Multivariate data analysis. a global perspective. Pearson Education, Upper Saddle River, NJ, USA, 816 pp.

Helgadottir, G., 2006. The culture of horsemanship and horse based tourism in Iceland. Current Issues in Tourism 9(6): 535-548.

Hung, K. and J.F. Petrick, 2011. Why do you cruise? Exploring the motivations for taking cruise holidays, and the construction of a cruising motivation scale. Tourism Management 32(2): 386-393.

Iso-Ahola, S.E., 1982. Toward a social psychological theory of tourism motivation. A rejoinder. Annals of Tourism Research 9(2): 256-262.

Iso-Ahola, S.E., 1983. Towards a social psychology of recreational travel. Leisure Studies 2(1): 45-56.

Iso-Ahola, S.E., 1989. Motivation for leisure. In: Jackson, E.L. and T.L. Burton (eds.) Understanding leisure and recreation: mapping the past, charting the future. Venture Publishing, State College, PA, USA, 653 pp.

Iso-Ahola, S.E. and J.R. Allen, 1982. The dynamics of leisure motivation. The effects of outcome on leisure needs. Research Quarterly for Exercise and Sport 53(2): 141-149.

Jang, S. and C.M.E. Wu, 2006. Seniors' travel motivation and the influential factors. An examination of Taiwanese seniors. Tourism Management 27(2): 306-316.

Jodd, S.J., 1994. Horse-based leisure activities and farm diversification. Anglia Polytechnic University, Cambridge, United Kingdom.

Jönsson, C. and D. Devonish, 2008. Does nationality, gender, and age affect travel motivation? A case of visitors to the Caribbean Island of Barbados. Journal of Travel and Tourism Marketing 25(3-4): 398-408.

Kaiser, H.F., 1970. A second generation little jiffy. Psychometrika 35(4): 401-415.

Keaveney, S.M., 2008. Equines and their human companions. Journal of Business Research 61(5): 444-454.

Kim, S.S. and B. Prideaux, 2005. Marketing implications arising from a comparative study of international pleasure tourist motivations and other travel-related characteristics of visitors to Korea. Tourism Management 26(3): 347-357.

Klish, E.M., 2009. The place of horseback riding in adolescent girls' development. Antioch University New England, Keene, NH, USA, 97 pp.

Kyle, G., J. Absher, W. Hammitt and J. Cavin, 2006. An examination of the motivation-involvement relationship. Leisure Sciences 28(5): 467-485.

Mak, A.H.N., K.K.F. Wong and R.C.Y. Chang, 2009. Health or self-indulgence? The motivations and characteristics of spa-goers. International Journal of Tourism Research 11(2): 185-199.

Maoz, D., 2007. Backpackers' motivations the role of culture and nationality. Annals of Tourism Research 34(1): 122-140.

Matthews, B. and L. Ross, 2010. Research methods. A practical guide for the social sciences. Pearson Education Canada, Newmarket, Canada, 520 pp.

Murray, E.J., 1964. Motivation and emotion. Prentice Hall, Englewood Cliffs, NJ, USA, 118 pp.

Newsome, D., A. Smith and S.A. Moore, 2008. Horse riding in protected areas. A critical review and implications for research and management. Current Issues in Tourism 11(2): 144-166.

Park, K.S., Y. Reisinger and C.S. Park, 2009. Visitors' motivation for attending theme parks in Orlando, Florida. Event Management 13(2): 83-101.

Pearce, P.L. and U.I. Lee, 2005. Developing the travel career approach to tourist motivation. Journal of Travel Research 43(3): 226-237.

Prebensen, N., K. Skallerud and J. Chen, 2010. Tourist motivation with sun and sand destinations satisfaction and the Wom-effect. Journal of Travel & Tourism Marketing 27(8): 858.

Rashid, M., 2011. Life lessons from a ranch horse. Skyhorse Publishing, New York, NY, USA, 185 pp.

Saad, G. and T. Gill, 2001. Gender differences when choosing between salary allocation options. Applied Economics Letters 8(8): 531-533.

Snepenger, D., J. King, E. Marshall and M. Uysal, 2006. Modelling Iso-Ahola's motivation theory in the tourism context. Journal of Travel Research 45(2): 140-149.

Thrane, C., 2000. Men, women, and leisure time. Scandinavian evidence of gender inequality. Leisure Sciences 22: 109-122.

Witter, R.F., 2000. Living with horsepower: life lessons learned from the horse. Swan Hill Press, Shrewsbury, United Kingdom, 268 pp.

Zimmer, Z., R.E. Brayley and M.S. Searle, 1995. Whether to go and where to go. Identification of important influences on seniors' decisions to travel. Journal of Travel Research 33(3): 3-10.

11. A comparative study into the impact of social media in the equine and agriculture industries

C. Martlew

Myerscough College, Bilsborrow, Preston, PR3 0RY, United Kingdom; c.f.martlew@gmail.com

Abstract

This study compared how equine and agriculture industry participants have adopted social media both in personal and professional capacities. Methodologies included online questionnaires, e-mail interviews and focus groups. Questionnaires highlighted significant associations in the use of social media, for example, in personal use equine participants were more likely to use Facebook while agriculture participants were more likely to use online forums; for business/professional use equestrians used social media sites more frequently at >10 times/day. Qualitative responses showed the majority felt positive about using social media sites for fast information sharing, improving communications with customers and broadening professional and social networks. The minority of negative opinions expressed that it is time consuming and the information is unreliable. Overall, the survey highlighted a stronger social presence by equestrians, while agricultural participants had a stronger professional presence and closer community networks.

Keywords: social media, equine, agriculture, comparison

Introduction

Social media is an online medium for communication and file sharing (Kaplan and Haenlein, 2010). The popularity and significance of social media has increased exponentially in the last decade (Qualman, 2013). Despite its beneficial effects; its use (and impact) in rural activities such as equestrianism and agriculture has not been established. Individual case studies highlight instances where social media has greatly influenced specific occurrences within the industries. For example, Sara Algotsson Ostholt, raised over £75,000 through Facebook to help keep her Olympic competition horse (White, 2012); the agriculture industry, meanwhile, was urged to embrace social media at the 2013 National Farmers Union (NFU) Annual Conference where, on Twitter, #NFU2013 trended worldwide (Wood, 2013). The aim of our study was to compare how equine and agriculture have embraced social media. The objectives were to survey a range of participants to determine the extent of their uses and perceptions of social media sites; compare the responses from the agriculture and equine industries and determine how social media has had a personal and professional impact.

Methodology

The methodology combined an online questionnaire for each sector; interviews via e-mail with specific individuals working in a communications role for their industry and focus groups with university students studying equine or agriculture. The questionnaires were distributed online and, most effectively, through social media sites such as Twitter; a tweet would be sent to an industry representative to share the link to the respective survey. These industry profiles had tens of thousands of followers, so the potential reach was extensive. The total responses to the online questionnaires were 407 for equine and 110 for agriculture; however, not all of those who responded answered the surveys in full. The difference in the number of respondents could be interpreted as being indicative of the activity of each industry members on social media. For the interviews, numerous industry professionals were canvassed but total respondents included 3 for equine and 2 for agriculture. The focus groups were attended by 4 equine students and 7 agriculture students. Utilizing mixed methods permitted a triangulated approach of quantitative and qualitative results and enabled a large amount of data to be obtained. The focus of the study remained the same across the survey range, but, with a variation in the mode of data collection any similarities in the results would provide more certainty of the conclusions, while any differences would provide further dimensions for consideration (Jick, 1979). Statistical analysis of the data included Chi-squared tests of association as well as Mann-Whitney and Kruskal-Wallis tests for comparison of rankings.

Results

The survey samples gained from the online questionnaires were obtained from a variety of industry participants. From both questionnaires the largest age group represented was 21 to 29 year olds, however, respondents overall ranged from younger than 17 to older than 60. The prevalence of young, working-age respondents could indicate a stronger online presence compared to other age groups which is supported by statistics that show the largest age demographic of Facebook users in the UK is currently 25 to 34 year olds followed by 18 to 24 year olds (McGrory, 2013). In comparison to equine respondents, the agriculture questionnaire generated greater participation from the 50 to 59 age group, which was, in fact, the second largest representation. This could reflect the nature of the farming industry where participants often remain for their entire lives, passing the business down from generation to generation. This contrasts with responses in the focus group that older participants were not receptive to internet applications but supports the activities shown by Driver (2012) and Wood (2013) that the industry is encouraging all participants to embrace social media. Thus, the age group of the respondents shows diversity in questionnaire participation and reflects some trends from wider UK statistics.

The roles of the questionnaire respondents show a variety of participation in the industries. Equine respondents described themselves predominantly as horse owners, while the majority of agriculture respondents stated that they were farmers. Due to the qualitative nature of the question, individual respondents could contribute to more than one category, so there were more answers than there were respondents and the 14 equine and 16 agriculture categories reflected a range of participants who contributed to the survey research. The fact that the biggest representation of equine respondents was horse owners and amateur riders reflects a more hobbyist/leisure contribution while the large proportion of farmers provided the agriculture responses with a more professional context. This accurately reflects the nature of the two industries; equestrianism can be viewed as more sport and leisure while farming is a livelihood and business. This means the two samples were potentially difficult to directly compare. The fact, however, that there was a section specifically for businesses and professionals from both industries to contribute provides a more equal platform from which to draw conclusions.

The disciplines and sectors in which respondents categorised themselves varied considerably. The top three disciplines chosen by equine participants were Eventing, Dressage and Show Jumping which, as Olympic sports, accurately reflects the main disciplines currently supported in the UK. Hunting and racing were the two most popular disciplines from the qualitative responses. These are two sectors where the British industry has a lot of history. A total of 26 categories reflect the variety of different areas which were represented. For agricultural respondents, the largest sector was crops followed by beef, sheep and dairy. Such sectors are some of the largest areas of the agriculture industry and a total of 20 categories of sectors were found. Overall, the different ages, roles and sectors of participants reflected a varied representation of the industries.

The sites most used by the different industry participants were similar, with Facebook being the most popular. Facebook was created in 2004 and, as such, is one of the oldest social media sites and, in the UK, over 50% of the population and 62.49% of the online population are members (McGrory, 2013). As such, it is unsurprising these two industries use Facebook the most. What is interesting is that, of the responses to the individual sites, equine participants were highly significantly more likely than agriculture to use Facebook, which supports a highly active presence on that popular site. They were also highly significantly more likely to use Pinterest which is one of the fastest growing sites in the UK (McGrory, 2013). Though the majority of agriculture respondents did show they use the main social media sites, they were significantly associated with forums and non-usage. This suggests that, when agriculture participants are online, they are embracing the mainstream sites less, preferring to use industry-specific forums and blogs. Conversely, equine respondents appear to be very active on the mainstream sites, in particular Facebook, following the trend of the wider UK population.

A similar pattern was found in the social media sites used in the business/professional section of the questionnaires. Again, equine participants were significantly associated with Facebook while agriculture, again, showed a preference for forums as listed in the qualitative responses. However, Twitter was one of the most preferred sites for agriculture participants which is supported by the findings from Driver (2012) and Wood (2013) that showed Twitter was seen as most useful site for professional networking. This suggests agriculture participants associate Twitter as a more useful business/professional social media resource while equine participants do not differ in their use of sites whether it is for personal or professional use.

The frequency with which industry participants visited social media sites for personal use was similar in both samples. The options that survey participants could choose from ranged from 'less than five times'; 'five to ten times' and 'more than ten times' which they then applied to either a 'per day' or 'per week' option. 'Less than five times per day' was the most popular frequency with which respondents visited their favourite sites, after which the responses followed the same sequence ending in 'less than five times per week' as the least chosen option. This shows similar trends for both industries with an overall tendency toward daily use of social media sites and, as such, revealing an active online presence. Of the individual frequencies, the 'more than ten times per day' option was significantly associated with equine respondents while the least frequent choice, 'less than five times per week', was highly significantly associated with agriculture respondents. This reflects a much more active online presence associated with the equine industry than with agriculture. The business/professional section showed more differences in the frequency of use. The equine participants showed a more active use of social media sites for their business/profession than their personal use as the highest frequency option, 'more than ten times a day', was the significantly preferred choice. The overall pattern of frequencies followed a less sequential trend than in personal use, though daily options were still more popular than weekly options revealing a tendency toward frequent use. In the agriculture sample, the most preferred option was the same in personal use at 'less than five times a day', as was the least preferred option of 'less than five times a week'. However, the other weekly options were chosen more than the daily options and, as such, reflects a more infrequent use of social media sites for business/professional purposes than for personal use and, overall, less activity than equestrian businesses/professionals.

The majority of both industry samples showed that they did follow or subscribe to industry-related groups or individuals on social media sites. A larger proportion of agriculture participants did not follow such groups or people, for which a variety of reasons were provided including: negativity, indifference, lack of awareness, no time and preferring to limit use to just friendship groups. Equine participants who also did not follow or subscribe to industry-related groups and individuals cited a lack of time, friendship connections only and negativity as their reasons. The negative points that were similar in both industries showed an aversion to the amount of information that following industry groups and figures would provide as well as viewing such information as unreliable. Meanwhile, further negative points made by agriculture participants included an aversion to conversing with other farmers in a social capacity, a lack of groups in a respondent's particular area of interest as well as one participant who described social media as intimidating, pervasive and mostly aimed at young people. These negative attitudes reflect trends that occurred throughout the questionnaire responses though were in a considerable minority.

When listing the industry-related groups and individuals that participants followed the responses were numerous and varied. Facebook and Twitter obtained the most responses and, as such, provided a better comparison between agriculture and equine samples. The categories of governing bodies, professionals/individuals, magazines/TV/other media and businesses were distinctly more popular than the others for both industry samples, which shows they preferred similar groups of sources for industry news and information. As equestrian business was the top category for equine respondents to follow on Facebook, this supports the preference for this site in business/professional use. On Twitter the equine participants mostly followed industry professionals/individuals which reflects the accessible nature of the Twitter site as anyone can follow a person's profile – whether they are a friend, stranger or celebrity. Governing bodies were the most-followed groups by agricultural participants on both Facebook and Twitter, which shows these official associations are a popular source of industry news and information. Agriculture respondents showed a greater preference for online communities (unofficial groups dedicated to certain areas of interest) than equine participants which supports the industry's closer association with forums as such groups similarly offer an area for individuals with a specific interest to share information. Overall, the agriculture participants' preference for online communities and associations shows a trend toward developing closer industry networks. The equine participants' preference for businesses and professionals reflects a trend toward a more commercial network in sourcing products for their horses as well as an interest in the sporting achievements of professionals.

The ranking of sources by the industry samples showed that both agriculture and equine participants, on average, rated websites as the most useful source of information. In listing most useful sources, the equine respondents then chose social media, followed by talking to other industry participants then magazines. Agriculture respondents ranked magazines second, then social media which was very closely followed by talking to others in the industry. Both industry samples rated journal articles as the least popular source. This shows that, on average, equine industry participants viewed social media as a more popular source of news and information than agriculture participants. The groupings of the sources revealed that equine participants ranked websites and social media significantly differently to their ranking of magazines and talking to others, and all were ranked significantly differently than journal articles. Agriculture participants, on the other hand, ranked websites and magazines significantly differently to social media and talking to others and all were, again, significantly different to journal articles. The fact that the two online sources were ranked highest by equine respondents shows a greater reliance and use of internet resources while agriculture preferred sources such as generation Web 1.0 websites and physical magazines. Moreover, the fact that agriculture ranked social media and talking to others so closely perhaps shows they view these sources as equal and supports the social media concept as a two-way communication tool, as well as agriculture participant's preference for forum use which is another medium for 'talking' to others to find out specific news and information.

The proportion of participants in the sample who stated they had their own online blog or forum was very similar for both industries and was in a minority. Those who did have their own blog or forum in the equine responses mostly used them for personal or discussion purposes, while from the agriculture responses business and discussion was cited as the most popular purpose. In addition, equine participants recorded mostly 100+ hits to their blog/forum, while agriculture respondents recorded mostly 1,000+ hits. The sample size is too small to make large scale assumptions. The responses, however, suggest equine and agriculture participants are somewhat equal in how many have their own blog/forum, but differ slightly in their purposes – equine participants are more inclined to use the blog for personal reasons while agricultural respondents show more business use.

With the use of forums agriculture participants showed more activity than equine respondents; agriculture respondents were highly significantly associated with visiting forums 'often', while the majority of equine participants visited forums 'sometimes'. When using forums the equine participants were significantly more likely to just view other people's comments, while agriculture respondents were significantly more likely to view *and* post comments on the forum sites. This reveals that the agriculture participants are more involved with online forums and blogs than equine participants. Moreover, the equine respondents had some negative comments about using forums including not trusting the knowledge of those that post comments. Overall, this tends to support Marston's (2013) report that there is some abuse and negativity around online forums. Agriculture participants, on the other hand, appear to have fewer negative experiences. This could be due to the closer sense of community that farmers appear to be creating online – indeed, from Driver (2012) and Wood (2013), it was shown that they experienced abuse and negativity from those *outside* the industry which could potentially motivate them to develop stronger networks online in order to 'stick together'. The forums the agriculture participants are visiting are industry specific (such as The Farming Forum and British Farming Forum) and offer a place for like-minded industry members to gather; this differs to mainstream social media sites which focus on allowing anyone anywhere to create and share a profile about themselves. Thus, agriculture participants perhaps value forums more and as such their online conduct may be more mannerly.

Overall positivity was expressed by both industry samples toward social media as a source of industry news and information and its continued use in the future, though equine participants were highly significantly associated with a fully positive response. Opinions expressed by agriculture participants in these qualitative responses support the theory that they are developing closer industry networks. Examples of such comments include: '…agriculture can be a lonely industry…so having social media helps to remain in constant contact with people working in similar situations'; '…agriculture needs to keep up with it to ensure sustainability and keep new/young people coming into the industry…'; '…it enables many people with shared interests and experiences to discuss and debate issues that matter to them'; 'it's a great way to keep abreast of events unfolding #sos dairy being a fantastic advantage…'. Two comments in particular demonstrate how social media provides more than just news and information as '…one can often find the sub story' and it '…flags up things which might otherwise be missed by conventional media'. Examples of more negative attitudes that show how agriculture participants favour forums, distrust some sites and the importance of social media supporting technology include: '…yes to forums and Twitter, trying to avoid using Facebook…'; 'I don't have an internet enabled phone yet though so it does restrict me somewhat' and '…remember to take everything with an extra pinch of salt as this is very much one person's opinion'. One particularly insightful comment shows one way the industry could improve access to industry-specific content online:

> it would be good to see more RSS feeds – boring when people go way off [on] Twitter and send lots of tweets about personal things of no interest to others. It would be good to see more sign posting/links to better social media in the agriculture industries – perhaps some awards to encourage higher standards.

Such comments support findings from Haley and Spedding (2013) and Wood (2013) that social media can have a positive effect on the industry by providing a place for sharing knowledge, information and news on current events and topics of interest. The negativity shown towards sites such as Facebook illustrates the trend that agriculture participants are less interested in the main sites and prefer to engage with others through forums.

Some interesting comments that represented positive attitudes from equine participants toward social media include: 'I find Twitter particularly useful at keeping me up to date with the latest news. Also following elite riders such as Laura Collett allows you to be almost part of their everyday lives…' and 'it is the best way to get up to date news in an industry that doesn't have traditional news outlets'. The minority of negative opinions was shown by such statements as: '…you don't always know who is behind the information. [I] prefer knowing who I am getting the information from…it [trusting information on social media] would be harder for people without experience' and '…there are a number of Facebook groups I have now left because the ideas they were sharing were unmonitored and encouraged bad practice. People are using social media to ask people questions they should be asking their vet'. Such opinions reflect how equine participants can use social media to build connections with the industry elite and receive news and information updates quickly and easily. The most common trend among those that thought negatively of social media was a general distrust of the quality of the information since the UGC (User Generated Content) nature of social media allows liberal speech. The comment above expresses concerns over the reliance some people have on such information – since searching out information online is so quick and easy, not to mention free, it is no surprise this has become a means of answering personal queries. It is, however, possible to fall victim to false information and poor advice, as demonstrated by Marston (2013). Thus, it is important users are aware of the risk and equestrian authorities may need to take on the responsibility of educating their members to lessen the misuse of social media sites.

Qualitative responses for the Business Use section were also predominantly positive with the majority of participants from both industries agreeing social media had benefited their businesses and that they will continue to use it in the future. Of the individual categories of responses, agriculture participants were significantly more associated with a positive and negative answer to using social media in the future for their business/profession which shows a cautiously optimistic attitude. Examples of responses that reflected positive attitudes among the agriculture participants' qualitative answers include: '…communicating through social media makes [farmers] appear friendlier and more professional to the public'; 'it helps customers know what is happening on the farm…'; '…[Twitter is] useful in explaining the day to day issues that we face…'; 'it encourages engagement and response to customers, good to have an online reactive presence' and '…We have gained a great amount of market position through our use of social media. Our brand has become more recognizable'. From the minority of participants that showed a negative reaction to social media in business, examples include: '…it puts us at risk for radical animal rights groups to target us'; '…I think it has decreased web traffic somewhat, which is a shame. Social media is impossible to search, except for Twitter' and 'a website is a good place to have a digital presence but blogs/social sites, etc. do not give the same impression generally, anything that gives people an opportunity to leave comments can damage your business and reputation and most people aim to sabotage and destroy with comments on topics they actually know very little about'. Statements that show a preference for forums and aversion to Facebook, in particular, include 'yes to forums and Twitter. Trying to keep minimal interaction with Facebook…'; 'Facebook is a crock and not very useful…if you aren't on it you are assumed not to be in business' and '…I cannot see any benefit in having a profile on social media for an agricultural business…however I will continue to use agricultural forums more'. Such responses show that agriculture participants have mostly experienced such benefits as better interactions with customers and promoting a more positive image of the industry, meanwhile issues include enabling abuse from industry outsiders, and that websites, Twitter and forums are more useful than Facebook.

Positive responses from equine participants which show that social media has provided benefits to business/professional use include: '…I have been able to spread the word of the business…helps me track people's interest in the business, reaches a wider audience for sharing news and offers, etc.…. I like the fact people can use Facebook to enquire about my services without ringing me, which may be daunting'; '…We have taken orders and made sales with the help of Facebook and Twitter…'; '…I have reached new clients in the media in different countries…'; '…it is where you will hear about new and innovative ideas, as well as keeping up with current trends in the equine industry. It is important to keep in contact with other professionals as well and social media is the easiest way to do so' and 'social media outlets are becoming far and away the easiest method to communicate information effectively and to a large audience'. Such comments reflect how social media has enabled equestrian businesses to communicate better with customers, increase sales, reach wider audiences and interact with other professionals. The following comments illustrate how important technology that supports social media connections is to equine participants: '… [It is] exceptionally useful to be able to research topics via phone from yard in cases of 'emergency;'' and '…people involved in the equine industry are very busy…the best way to catch them is through social media since smart phones are the way of the future. People use social media to touch base with their clientele and set appointments which is a huge benefit to the equine industry since they can respond quickly through their smart phones'. Some participants show insights into how, with more knowledge and awareness, social media can provide further benefits to the equine industry: '…When used correctly, social media is a great tool to increase brand awareness internationally for little money. The trick is learning how to turn brand awareness into sales' and '…with more horse related groups interconnected…think about how fast information could be distributed about stolen horses…a fast spreading virus…someone in need of a specialist…the possibilities are endless'.

The minority of negative comments were reflected in statements such as: 'it could also give negative feedback that could damage the reputation of your business'; 'I find everything I need for my profession through websites and contacting people/organisations'; '…advertising on Facebook is unprofessional…'; '…people need to be aware that they are not getting the full picture of what is going on. They are getting snippets; social media is not a reliable news source because of its lack of organization, which is also probably its strength'; 'I use it more as a personal…network. If I need information I usually go to the source or websites to get it' and '[I am] more likely to use websites. Social media is very time consuming'. One negative comment that was unique but interestingly similar to difficulties experienced by agriculture participants was '…if you have strong opinions about an area of interest to radical animal rights activists, you will get negative feedback'. Such comments demonstrate how equestrian participants have experienced many benefits from incorporating social media into their business or profession including improving brand awareness, access to information and audience reach. However, the few negative responses reveal it is sometimes seen as unreliable, unprofessional, time consuming and allows negative feedback to be made public. Overall, equine and agriculture participants are similarly aware of the benefits of using social media for business, however, agriculture responses show more awareness of how it benefits the whole industry, whereas equine seem mostly conscious of the individual's advantages.

Discussion

The interviews reflected opinions of participants in a professional role who are in a position to observe the effect of social media on organisations and businesses across the industry. The agriculture interviews supports Katims (2010), Durban (2011) and Wood (2013) in that the industry wants to use social media for promoting a positive industry image, connecting isolated farmers and spreading campaigns. The positive responses in the questionnaires are reflected in how interviewees have seen social media benefiting businesses in the industry by providing a means of connecting producers to their consumers. More similarities were shown in Sharon Hockley's reference to anti-campaign protestors being able to voice their opinions against some industry practices through social media. Added to this is the need for improved technological connections to allow farmers in rural areas to access online media which could, in part, explain the relatively slower growth in participation. The NFU revealed a desire to improve the education, awareness and participation of farmers who do not perceive social media as a tool for them – the existence of such participants is shown in the negative responses in the online questionnaires. From the interview with Farmers Weekly's community and farm-life editor, Tim Relf, there was a view that the industry was readily embracing social media and participants, contrary to stereotypes, commonly incorporate modern technology into their practices. In addition, Relf identifies the strong community connections that farmers online are developing which is seen through their prevalent forum use. Overall, the interviews support much of what was found in the literature that there is a desire to better utilize social media for the benefit of the whole industry, as well as illustrating themes from the questionnaire in the popularity of forums and negativity from antagonists.

The interviews with equine participants showed a consensus of opinion that social media provides many benefits to businesses and organisations including providing a platform for digital marketing, enabling customers and members to keep abreast of events and connecting participants globally. The interviewees also identified that the main negatives are that members of the industry who misuse social media by sharing information could potentially damage reputations. This is supported in the responses to the online questionnaires where those who thought negatively of social media mistrusted the reliability of some of the information shared.

Moreover, Lucy Hurst from British Dressage described how social media can be time consuming but is important enough to require its development as a professional skill. Some respondents in the online questionnaires also thought of social media as too time consuming, however, the response from interviewees showed that the benefits can be worth the time spent developing an online presence. The knowledge and awareness in how to establish a successful profile online could be lacking in some equine participants and some encouragement is perhaps needed for better collaboration and increased awareness to use social media more effectively. From the interview with Liam Killen (director of the Equestrian Social Media Awards), he describes the industry as almost separatist as it is so different to other industries, however, social media can be used to break down such barriers and it is important the industry keep up to speed with social media as it affects the world of communications. The three interviewees all expressed intentions to remain in the world of social media and encourage its development across the industry. The fact that, in the online questionnaires, positivity toward social media was felt by the majority of participants shows it is largely being embraced by the industry. Overall, social media appears to be highly successful as a personal communication tool, though some better awareness is needed among participants to discourage certain behaviours that are damaging the industry's reputation while continuing to benefit its businesses and organisations.

Finally, the Focus Group participants represented the younger generation of up-and-coming professionals who were beginning their integration into various roles in the industry and some of their responses highlighted conflicts with other findings. Responses in the equine focus groups illustrate some awareness of social media as beneficial for businesses, however, the overall attitude and usage reflected a tendency toward personal use only and not for building professional networks. This can be demonstrated by their preference for Facebook which is designed more for friendship and personal connections and negativity toward Twitter which allows connections to anyone, professional or personal. As non-professional participants it is perhaps understandable their view of social media was limited to informal social interaction, however, as university students their careers would be starting relatively soon and would perhaps benefit from being instructed in how to create a professional online presence. After all, social media has become a recruitment tool and its wide accessibility allows people to view anything (good and bad) that people share – the students showed their awareness of this fact as they spoke of privacy settings. However, it is perhaps this lack of knowledge in how to use social media for professional benefits that has caused their aversion to building a larger online presence through the different sites. If the positives felt by the interviewees could be shared with participants like these students then they would be better equipped to enter the professional sector of the equine industry with a positive online presence.

The responses in the agriculture student focus group similarly showed minimal awareness of social media as a professional development tool and Facebook was, again, the most used site. Indeed, some negativity was shown toward Twitter, though this was born more out of the lack of knowledge on how to use it. Indeed, when asked if the students had heard of AgriChatUK the majority had no awareness of its existence. They also expressed doubt in usefulness as a way of communicating with other participants, citing older age and a stereotypical aversion to the internet as a reason farmers did not participate online. This is somewhat supported by the interviews and literature in that the industry needs to actively encourage traditional farmers to embrace social media. The second largest group of respondents to the *online* questionnaires, however, were 50 to 59 year olds which shows some success in the push to get later working age farmers to participate online. Thus, the lack of knowledge in these students is perhaps indicative of the affect stereotyping has had within the industry itself – that it is assumed younger people are more technologically adept and, as such, are not being targeted by those who wish to encourage professional development in online use. This is similar to the findings of the equine focus group and perhaps shows a gap that both industries are missing out on – that their younger generations still need 'instructing' in how to use social media as more than just personal communication but as a means of developing professional networks.

Conclusions

Overall the equine industry shows a strong online presence on a variety of sites and views social media as source for industry news, sharing information and connecting with others but perhaps in a more conservative way. The benefits of social media as more than a personal communication tool is well known by some current professionals and is something those participants hope will benefit the whole industry. Some gaps in awareness and knowledge do exist among some participants and as such could potentially lead to misuse. Indeed, the equine industry appears to be condemning itself, as most negativity is felt to come from within. This indicates perhaps some weakness in the sense of community that equestrians have online. Moreover, if younger generations of participants are unmotivated to develop professional online networks to assist in their career, this weakness in industry collaboration online may continue. As such, the industry may need to focus some resources into developing better education and awareness in participants; organisations such as the ESMAs demonstrate an active endeavour toward this.

For agriculture participants, the young to middle-aged professionals seem to be fore-runners of social media use, while the much younger and older farming participants seem less aware of its professional benefits. This could perhaps indicate why, when the industry *does* use social media, it is more for business and industry-specific purposes as opposed to just another social communication tool. Moreover, there appears to be a strong sense of community in those farmers using social media. Indeed, negative experiences were more often felt from users outside the industry through stereotyping and protesting, which could encourage a more defined identity in those farmers who are online. This does not, however, suggest the industry is behaving in an insular fashion, as participants expressed a desire to connect with their consumers and promote a better image of the industry to outsiders. This pro-active use of social media is encouraging. If younger generations are showing some lack of awareness in this area, however, the industry may need to target those users to ensure progression.

In answer to the research question, there is some variety in the uses and perceptions of social media in the equine and agriculture industries with similarities and significant differences in both personal and professional capacities. General trends and patterns throughout the literature and surveys have revealed fascinating insights into the behaviours of industry participants. Equestrians appear to have a stronger social online presence with a steady growth in professional and business use. In agriculture, though participants are perhaps slower than equine in embracing social media, they seem to have a stronger professional presence and closer community networks. Both industries look to be encouraging more usage in the future though both could ensure this happens by focusing further on the younger industry members who are beginning their careers. In addition, while agriculture participants are attempting to address negativity from outsiders by promoting a positive industry image, the equine industry must target its own members in reducing the negative use of social media which appears to be deterring participants from creating an online presence and could potentially be off-putting to outsiders. Social media is continually growing, developing and shaping people's lives and, as technology advances, the ease of access to such online resources will only increase. The significance of this means the potential for improvements in communication is considerable and cannot be ignored.

References

Driver, A., 2012. The year farming embraced social media. Available at: http://tinyurl.com/px6ny6c.

Durban, E., 2011. Agriculture finding its voice on social media. Available at: http://tinyurl.com/o62a8h6.

Haley, S. and A. Spedding, 2013. AgriChatUK – 'Innovation in farming'. Available at: http://tinyurl.com/ofwvgzy.

Jick, T.D., 1979. Mixing qualitative and quantitative methods: triangulation in action. Administrative Science Quarterly 24(4): 602.

Kaplan, A.M. and M. Haenlein, 2010. Users of the world, unite! The challenges and opportunities of social media. Business Horizons 53: 59-68.

Katims, L., 2010. Farmers milking social media to promote agriculture. Available at: http://tinyurl.com/nuww8sg.

McGrory, M., 2013. UK social media statistics for 2013. Available at: http://tinyurl.com/c8rupmj.

Marston, V., 2013. 'Equestrian forums: priceless resources or potential danger?'. Horse and Hound, January: 60.

Qualman, E., 2013. Social media video 2013. Available at: http://www.socialnomics.net/2013/01/01/social-media-video-2013/.

White, C., 2012. Fans rally to buy 'silver sara's' next top horse. Horse and Hound, November: 7.

Wood, R., 2013. Agricultural industry urged to embrace social media. Available at: http://tinyurl.com/osde4nv.

12. The acceptance of complementary therapies amongst equine communities and what therapists need to know

J. McKeown

University of Aberystwyth, Planning Office, Penglais Campus, Aberystwyth, SY23 3BF, United Kingdom; jum1@aber.ac.uk

Abstract

There is a widespread growth of complementary and alternate therapies (CAM) amongst the equine community today, possibly fostered by popular equestrian magazines and also the growth of Natural Horsemanship. The growth of therapies implies a growth of therapists, who make their living from treating horses and need to understand the different equestrian sub-cultures in order to target their services successfully. This research looks at the consumption of CAM within horse owners in Wales and the implications of the findings for therapists working in this area. The therapists are often female entrepreneurs based in low income rural areas with no other major source of income.

Keywords: CAM, equestrian, sub-cultures, rural economy

Introduction

The term 'Complementary and Alternative Therapies' (CAM) is now being used by many authors where in the past the terms complementary and alternative medicine were used synonymously in error (Botting and Cook, 2000) (Alternative therapies replace conventional medicine and complementary medicine works alongside). There are many different types of therapies for horses, from acupuncture which has to be administered by a vet, to homeopathy, chiropractic, sports massage and osteopathy as well as more esoteric therapies such as Reiki, Bowen or crystals. With such a choice, how can the average horse owner know which therapy to use and what will bring the best relief for his/her horse? What are the implications for the therapists themselves who are trying to earn a living from administering these treatments?

As with other sub-cultures, opinion leaders are important for the adoption and diffusion of new products – including CAM (Chan and Misra, 1990). The equestrian lifestyle is varied but basically split into two main areas in the UK – Traditional and Natural Horsemanship. The former features horses kept in stables; ridden with bits, whips and spurs; shod and fed concentrate feeds (Rose, 1977). The latter has horses kept outside in herds; ridden with bitless bridles and no other 'aids'; the horses have no shoes on and are fed fibre based food (Roberts, 2002). Many owners keep their horses in a natural way but ride with traditional equipment (McKeown, unpublished data), mixing the two approaches.

The research

Fragmentation in postmodern society has resulted in the creation of new types of sub-cultures (Cova and Cova, 2002; Firat and Dholakia, 2006) which create their own 'consumption spaces' (Haslop *et al.*, 1998; Shields, 1996, p. 101). For example, in the equestrian sub-culture these include livery yards, saddlers and Events where members of the communities can meet and consume equestrianism. With the advent of the internet, this is moving to online forums and social media's extended consumption spaces such as facebook groups. These tend to be oriented around rituals of exclusion, inclusion and group membership (Shields, 1996, p. 108) and offer ways to extend knowledge boundaries within related groups. Birke (2007) offers an example of how the Natural Horsemanship culture change has a lot in common with CAM and changes within horsemanship. The use of barefoot trimmers instead of traditional farriery can be seen as an instance of this.

This research was undertaken in 2011 and 2012 and attempts to understand the marketing issues faced by equine complementary therapists and offer recommendations to them. It began with an introspective look at equine subcultures. Syncretic forms of introspection were adopted, where the researcher and respondents were used as data with little differentiation made between the two during analysis (Wallendorf and Brucks, 1993) (The author is an equestrian and equine shiatsu practitioner). The 173 respondents from all regions in Wales, shared factual data via questionnaires, and a subset of 16 discussed deeper insights towards their horses, equestrian lifestyle and CAM during in-depth interviews. Four of the interviewees were CAM practitioners and 3 were vets (two of whom also practiced CAM). The remainder of the interviewees were horse owners. The questionnaire was constructed after in-depth analysis of an introspective piece written by the researcher which allowed deeper insights and questions to come to the fore (Ellis, 1999).

Findings

The preliminary findings from respondents show that although some of the 21 therapies specifically mentioned in the research are so well-accepted that they are thought of as mainstream (chiropractic and acupuncture for example), others are still regarded with suspicion (such as crystal healing) and whilst herbs are widely accepted, zoo pharmacognosy (oils made from herbs and other natural compounds which the horse chooses himself) is not despite being mainly herbal (see Table 1 for therapies).

Mixed and traditional equestrian lifestyles amounted to 87% of respondents (43 and 44%, respectively) and the remaining 13% practised Natural Horsemanship. The effects of opinion leaders within these

Table 1. Popularity of different forms of CAM (n=173).[1]

	Do use horse	Would use horse	Do use me	Would use me	Would not use	Don't know	Not heard of	Total
Acupuncture	6	70	29	50	28	20	0	203
Alexander	5	32	12	51	5	33	34	172
Aromatherapy	24	47	38	52	15	17	1	194
Bach flower	29	39	47	29	13	14	30	201
Behaviour	18	50	12	28	16	38	2	164
Bowen	20	32	17	26	9	29	44	177
Chiropractic	48	62	37	56	8	15	2	228
Crystals	9	15	11	15	49*	38	17	154
Cymatics	5	6	3	7	9	19	98*	147
Herbs	78*	45	35	44	9	13	1	225
Homeopathy	36	59	39	47	12	28	5	226
Lasers	9	41	2	36	20	52*	3	163
Magnets	35	63	24	40	10	27	4	203
Massage	58	77*	58*	59*	3	6	0	261
Osteopathy	33	63	34	52	5	25	4	216
Podiatry	16	27	10	37	9	38	25	162
Reiki	29	45	31	35	18	31	9	198
Shamanic	2	14	5	12	15	24	79	151
Shiatsu	15	51	16	44	11	35	21	193
TTT/TFT	5	24	6	15	7	10	87	154
Ultrasound	10	58	14	44	9	39	9	183

[1] Numbers marked with an * are the highest for that column.

communities cannot be overestimated with many respondents citing use and purchase of CAM via recommendations from those they trust on their yard or within their equine circle. This concurs with the findings of Chan and Misra's 1990 study on opinion leaders.

Demographics and lifestyle of respondents

As with many other studies on equestrian culture, the majority of the respondents were women (see Birke and Brandt, 2009 and Kay, 2008 for example), in this case it was 91%. This female dominance affects purchasing decisions, especially in the CAM world where women are seen to be more in favour of these therapies than men (see Ernst and White, 2000 for example). The female respondents also tended to ride for pleasure whereas the males competed or hunted as well as using their horses as employment (a horse logger for example).

Of the 173 questionnaire respondents, thirty-one worked in the equine industry with the rest of the respondents having a variety of roles from business managers to farmers and students, to staying at home. 54% s of respondents were aged between 41 and 60, and 21% aged between 21 and 30. There were 19 over 60, 3 of whom were over 70, and 9 who were 20 and under. Seventy five people had income of between £15,000 and £45,000 with 44 respondents earning over this amount (the UK average is £26,000). The respondents all lived in Wales and the English border counties.

From a choice of 9 categories, 110 riders included hacking and fun rides as part of their equestrian lifestyle. Only 9 people drove their horses. The number of riding club/pony club (52) and local competitors (55) were similar (22 of whom did both) but only 29 out of the 171 competed at national or county level in their discipline. Twenty four were parents of children in local riding or pony clubs and 25 rode in a riding school, with or without their own horse. Thirty three people said that they hunted with horses.

There were a total of 676 horses kept between the respondents, 80 were kept in livery and the rest (596) were kept at home. According to 2010 research by Horse & Hound, there are around 65,000 horses in Wales and another 56,000 in the Welsh border counties (Horse and Hound, 2010).

Twenty nine respondents kept their horses at livery (an average of 2.7 horses per respondent) and 134 kept theirs at home (an average of 5.1 horses per respondent). Two respondents owned 30 horses (one claiming they were in livery), and the owner of a riding school had 53, but the majority of people owned between 1 and 5 horses. Those with horses in livery mainly had only one or two horses (when removing the outlying respondent with 30 in livery which skewed the results).

Only two of the seven respondents who owned ten or more horses were equine professionals. These were an owner of a riding school (53 horses) and a livery yard owner (11 horses owned plus 19 liveries). There were 2 farmers (30 horses and 12 horses) and the others were in office or professional work.

Regardless of the number of horses owned by people and their income, the majority of owners (104) spent between £1000 and £6,000 a year on their equine activities – which included horse keep, purchase of tack and other items, and competitions, travel, etc. One person spent between £15,000 and £20,000 a year but she was an equine assisted coach so her nine horses were her business. She used therapies on them regularly. Of the 2 people who spent over £20,000, one was a male racehorse trainer who had 4 horses of his own and never used CAM and the other was a female owner of a riding school with 53 horses and who didn't use CAM on them but did occasionally on herself. Six people didn't want to add up what their horses cost them a year.

Purchase behaviour varied, but an overall majority (145) still use retailers for their equine purchases and 57 said retailers were their first choice of purchase channel. EBay (98), other internet sites (116) and catalogues (95) were the other main channels used to buy, with other internet sites more popular than EBay overall. Magazines were the least popular choice for purchase.

There was a wide variety of horse types kept by respondents, some more specialist than others – for example the sports horses bred for jumping and eventing, and the warmbloods for dressage. Unsurprisingly the majority of horses were Welsh ponies and cobs – either pure bred or having some Welsh blood in them. There were also other breeds of horse that were crossed. After the Welsh (82 people owning them), the most popular bloodlines were Thoroughbred with 63 and Warmblood with 49 owners. Some owners had a variety of horses, for example one owner had 4 Arabs, 1 Thoroughbred, 1 Dartmoor pony, 1 donkey and 1 miniature horse.

Osgood's semantic differential scales (2009) were employed to understand owners' attitudes towards their horses and how they were valued (Friends, to Status symbols; Partnership to, A way of competing; Members of the family, to Possessions; Close bonds, to No feelings). In the first category, only one person saw their horse as a status symbol with the majority seeing their horses as friends. Partnership was more varied but again, people felt their horse(s) were partners with only 3 seeing them as a way of getting out or competing. The third category echoed the first one with the majority of people seeing their horses as members of the family and only 2 saw them as possessions. Finally the majority of people had close bonds with their horses and only 2 had no feelings for them – both were men.

General horse care

The traditional way of keeping horses is stabled at night with turnout in the day – either for a short time or for most of the day. As horses are herd creatures most owners try to turn out with one companion or within a small herd. There are a number of owners (78) who keep their horses out throughout the year and a further 36 who keep them out all summer. This is a much more natural environment for horses and 17 owners also kept their horses in a barn as a herd instead of individual stables. Only 13 owners kept their horses on their own in no sight of others. Thirty two of the owners who stable their horses choose to let them out at night in summer as there are fewer flies to bother them.

All respondents had their horses' feet attended to but at different time intervals. Six to eight weeks is the accepted time for having shoes replaced and this was confirmed by the high number of respondents reporting this (135) with 5 others saying they had their horses shod every 5 weeks. Trimming is often done only every 3 months or so and 15 respondents gave this answer. The remaining respondents had their horses feet attended to at longer intervals or 'as and when'.

Traditionally horses are wormed 3 or 4 times a year (Rose, 1977, p. 192) and 73 respondents reported doing this. Twelve wormed more regularly (every 6 to 8 weeks) and these were fairly new to keeping horses. Others wormed at greater intervals with 27 using intelligent worming methods that include taking egg counts in droppings before worming. Only 1 respondent used a herbal wormer.

Equine dentistry has grown as a practice in its own right for the past forty years or so and now most owners have their horses teeth checked. Only 8 respondents didn't have this done. Most (109) had them checked annually and a further 22 had them checked every 6 months. Two had teeth done every 2 years and 19 as required or occasionally. The remainder checked them themselves or had the vet to do it.

Having saddles checked is something newer and with greater understanding of back issues more owners are taking this area more seriously. Only 12 owners never checked their saddles, but there was greater variation in answers in this category than the others. Five checked every 6 to 8 weeks, 7 every 3-4 months, 20 every 6 months and 53 annually. Of the remainder answers varied from 'every time I ride' to 'horses not ridden'. It appears that many owners check the saddles themselves and are unlikely to be qualified in this area.

Complementary therapy use

Of the overall 173 respondents, twenty one stated they delivered complementary therapies and 149 did not although 5 of the latter had some form of training in CAM and treated their own horses. Three declined to answer. Of the 21 therapists, 10 treated both people and horses, 6 treated people, 2 treated people and their own horses, 1 only treated horses and 2 declined to answer. There are a variety of therapies delivered but the most popular were reiki (11) and massage (5). There were 20 other therapies mentioned with one or two people practising each, ranging from counselling to homeopathy. Note that some therapies require many years of training (akin to undergraduate medical school) whereas others can be learned and applied after a few weekend seminars (Pietroni, 1992) and at present in the UK regulation is voluntary within disciplines.

In total 105 people had CAM treatments on themselves, with the majority of these having treatments when needed (36), between 2 and 6 a year (24) or rarely (28). Only 18 people had treatments more frequently than bi-monthly. Sixty five people never had CAM treatments.

A variety of reasons were cited for equestrians using CAM on themselves and unsurprisingly much of this was to do with musculo skeletal issues (31 answers) and pain relief (13). Riding is accepted as one of the most dangerous sports and therefore both rider and equine injuries are common. All riders need good posture and balance but none more so than dressage riders. However only one of the respondents who said that they get benefits of better posture and balance from CAM was a dressage rider.

There were 109 respondents who used CAM on their horses with the majority (61) having occasional treatments only. Only 23 had regular treatments and another 25 when needed. Fifty people never used CAM on their horses. However when given the opportunity to say what particular therapies they did or would use later in the questionnaire many ticked that they would or did use some. This could be due to subjectivity in what constitutes a complementary therapy. Herbs, for example, are often considered to be a feed supplement.

When attempting to understand how people get to know about complementary therapies in order to use them, seventy six of the respondents got their information from friends. This concurs with the current understanding that word of mouth is one of the most powerful forms of marketing (Trusov *et al.*, 2009). Thirty two used articles as information sources and after that there was little to choose between the other categories (between 20 and 23 in each) such as leaflets and demonstrations. Other answers included other complementary therapy practitioners, growing up with it, wellbeing fairs, associations, working for practitioners and the doctor.

Reasons for choosing CAM

Of the many reasons for choosing complementary therapies over allopathic medicine, 'advice from friends' was the third most popular reason, again confirming the power of word of mouth (when adding in the responses saying 'my trainer uses it' this makes word of mouth even more powerful, further concurring with the Chan and Misra, 1990 study). The most popular answer for why the choice was 'keeping horses well,' which resonates with the understanding that CAM is about wellness and holistic healing rather than illness. Fifty four of the respondents said that they believe in CAM and 49 believe it treats problems rather than symptoms.

Forty two people said they use CAM because it is non-toxic, however it is well known that some herbs used in large quantities can be problematic and just because some things are 'natural' it doesn't mean that they aren't harmful. For example, hemlock was used as a poison for millennia and is still known today as one of the most poisonous plants growing in the UK. One herb (valerian) which is used in many herbal supplements was put on the banned substance list (equine doping) in 2003 due to the strength of its calming properties (Russell, 2003).

Other popular answers to this question included the enjoyment of the treatments by the horses; good results; a holistic approach; and a belief in CAM. People by and large still had faith in the veterinarian although some said that using CAM was cheaper. This shows a complementary rather than alternative approach towards CAM by most horse owners who use it.

Few respondents said that they used CAM through reading articles (32), adverts (20), leaflets (23) and demos (21) – so traditional marketing techniques were the least useful in persuading people to use CAM. Friends using CAM was by far the biggest persuader with 51 people choosing this option.

Thirty nine people said that they use CAM as it gives good results. These were broken down into relaxation, lameness relief and suppleness with 7 responses each; holistic improvement (6); muscular (4); pain relief and relief of condition with 3 responses each; sarcoids, back and traditional not working 2 responses each and with one response each; behaviour, no drugs, decrease of recovery time, cough, horse enjoys and ownership of my horse's treatment. (Some respondents chose two or more reasons for gaining good results) and one respondent used CAM as they were not sure what to use to treat the condition. Overall people looked to opinion leaders (friends) to start using CAM and then carried on due to the results obtained.

Frequency of use of CAM

Twenty one different therapies were outlined in the questionnaire, all of which can be used on horses and humans although often delivered differently e.g. podiatry. Some were the 5 most popular therapies (acupuncture, chiropractic, herbs, homeopathy and osteopathy) taken from various reports (such as: Fulder and Munro, 1985; Ong and Banks, 2003) and others were lesser known, such as cymatics (sound and light therapy), shamanic and thought field therapy (TFT). The rest were therapies that were more popularised but did not come in the top 5 list such as reiki and also those that may be considered mainstream now such as ultrasound and massage (Table 1).

There were 162 people who would or do use some form of CAM on themselves or their horses, 71 who would not use 1 or more forms of CAM and 130 who answered that they weren't sure or had not heard of some forms of CAM. Overall there were only 5 respondents who said they did not or would not use CAM on themselves or their horses.

From the data about therapies used (Table 1, popularity of therapies chosen) there is a trend showing that the most popular number of therapies that would be or are used by respondents is between 8 and 14. Therefore if people are going to use CAM then they are likely to adopt a range of treatments rather than just one or two.

Overall most of the respondents were either completely open to some form of CAM (i.e. they either would or do use it – with 3 saying they would or do use all of the forms of CAM in the table), or needed some further input to be open (i.e. were not sure or hadn't heard of some therapies) and only 2 would not use any form of CAM at all. These responses were taken from the table of 21 therapies and often conflicted with the responses in the main part of the questionnaire where 65 people said they never used CAM on themselves and 50 said they never used CAM on their horses.

It would appear that the term 'complementary or alternative therapy' has different connotations to some people than others. For example one person interviewed who was adamant that he would never use and never had used CAM, responded in the table that he regularly fed his horses herbs and only responded that he would never use one form of CAM (chiropractic), answering 'don't know' to the remainder of therapies in the table. This subjectivity highlights one of the difficulties that therapists may have in getting their therapies taken seriously.

Of the 2 people who would never use CAM (or the 21 therapies noted in Table 1), both kept 1 horse at home in a traditional manner. One was a teenager, the daughter of a farmer and the other an older woman who had owned horses for over 60 years. Both respondents bought only from retailers, thought of their horse as part of the family and had a close bond with them.

Two people said that there were 15 therapies they wouldn't use and both said that there were 6 they would use – therefore no 'don't know' or 'not heard of' answers. Both had said they would not use cymatics which is a very uncommon form of therapy that 98 of the respondents said they had not heard of. Although it cannot be confirmed it is likely that these respondents answered that they would not use any therapy that they were either not comfortable to use or unaware of. This confirms the necessity of generating an awareness and understanding of therapies if they are to be trialled and adopted. Both of the respondents were equine professionals.

Of the 3 people saying that they would or do use all forms of CAM outlined; all were female and in the age range 41-60. All had owned horses for between 10 and 18 years and felt that their horses were part of the family and there was a strong bond with them. None of the women were CAM therapists.

There were 9 people who would try or do use 19 or 20 of the 21 therapies listed and a further 16 who would use 17 or 18 of them. Of the nine, 4 were therapists (including one of the men) and 1 worked for an holistic vet. There were 7 women and 2 men; 7 keeping their horses in a mixed environment, 1 traditionally and one natural horseman. Six kept their horses at home, 2 in livery and one person didn't own a horse. Six were in the age range 41-60.

Cymatics (sound and light therapy) was the least popular therapy and 98 people had never heard of it. When removing this category crystals was the least popular with 49 people saying they would not use it. The most popular was massage overall, having the highest number of people saying they would use on their horse, do or would use on themselves (Overall popularity is the sum of the 'would' and 'do' categories for horse and person in the table above). Massage was the most popular therapy that people use on their horses with 58 people saying they do use it and another 77 who say that they would use it, followed by herbs (78 + 45) and chiropractic (48 + 62). Seventy respondents said that they would use acupuncture on their horse but only 6 already do. However 28 people would not use acupuncture and 20 weren't sure.

In total there were 920 responses indicating that some form of therapies would be used on horses and 779 that said they would use one or more therapies on themselves. There were 490 responses that showed some therapies were used on horses and slightly fewer (480) used on the owners. There were only 280 responses of the total (3975) that indicated that therapies would not be used. There were 475 responses for therapies that had not been heard of, over a fifth of these for cymatics, 87 for ttt/tft and a further 79 for shamanic, showing that the majority of the therapies in the table were known about to a greater or lesser degree. However 551 responses showed that people had to be convinced about using particular therapies with lasers having the highest number of this response (52) and ultrasound second with 39 responses. Equal third were 'behaviour', podiatry and crystals with 38 each.

'Behaviour' and podiatry are fairly new in equine terms and are strongly linked to the natural horsemanship community who expect their horses to go barefoot and work with the owner in a variety of ways. In the traditional community horses are shod and are expected to conform to the rider's wishes. As the majority of respondents considered themselves to be traditional or using mixed methods of horsemanship, these responses are understandable. Both ultrasound and lasers are widely accepted technologies today in certain areas but appear not to have total acceptance from people for use in healing rather than scanning and as light carriers – although laser surgery for improving vision is gaining momentum.

Recommendations for practitioners

Having analysed the data, early recommendations for therapists include being well-qualified; professional and confident; experienced around horses and able to educate owners to remove anxieties rather than marketing to them. These observations are based on comments made by respondents, for example: linking their area of therapies to one or more that is widely accepted or using local case studies and opinion leaders within their target equestrian sub-culture. Because many people use a number of therapies partnering with others to form networks offering a variety of treatments may be beneficial. Showing middle-aged female grass roots riders that there is an empathetic understanding of what their horse(s) mean to them could encourage adoption of therapies as this demographic is the most likely to try new forms of CAM (see findings and Oldendick *et al.*'s 2000 study). Development of articles explaining their therapies, linked to case studies, posted on forums and published in the equestrian press (if possible) could widen the acceptance and educate the therapists' target markets. Being part of the equestrian sub-culture as a horse owner and/or rider will also encourage acceptance of their therapy.

Conclusion

As equine communities move from traditional to mixed environments for their horsemanship, they are more open to adopting CAM. However, the research also shows that they not so open to adopting full Natural Horsemanship methods, but instead adapt Traditional and Natural Horsemanship to suit their own equestrian lifestyles. For the majority of horse owners in Wales, horses are no longer used for work, and in most of the cases here are seen as part of the family. They are very much a leisure rather than a working animal – almost a pet at times for this particular equine sub-culture. Equine CAM therapists should be mindful of this when marketing to their client base, who are mainly middle aged women enjoying riding club activities with their horses.

References

Birke, L., 2007. Learning to speak horse: the culture of 'natural horsemanship. Society and Animals 15: 217-239.
Birke, L., and K. Brandt, 2009. Mutual corporeality: gender and human/horse relationships. Women's Studies International Forum 32: 189-197.
Botting, D.A. and R. Cook, 2000. Complementary medicine: knowledge, use and attitudes of doctors. Complementary therapies in Nursing and Midwifery 6: 41-47.
Chan, K.K. and S. Misra, 1990. Characteristics of the opinion leader: a new dimension. Journal of Advertising 19(3): 53-60.
Cova, B. and V. Cova, 2002. Tribal marketing: the tribalisation of society and its impact on the conduct of marketing. European Journal of Marketing 36(5/6): 595-620.
Ellis, C., 1999. Heartful autoethnography. Qualitative Health Research 9: 669-683.
Ernst, E. and A. White, 2000. The BBC survey of complementary medicine use in the UK. Complementary Therapies in Medicine 8: 32-36.
Firat, A.F. and N. Dholakia, 2006. Theoretical and philosophical implications of postmodern debates: some challenges to modern marketing. Marketing Theory 6(2): 123-162.

Fulder, S.J. and R.E. Munro, 1985. Complementary medicine in the United Kingdom: patients, practitioners and consultations. Lancet 2: 542-545.

Haslop, C., H. Hill and R. Schmidt, 1998. The gay lifestyle – spaces for a subculture of consumption. Marketing Intelligence and Planning 16(5): 318-326.

Horse and Hound, 2010. How horsey is your county? July, IPC Media, London, United Kingdom.

Kay, J., 2008. A blinkered approach? Attitudes towards children and young people in British horseracing and equestrian sport. Available at: http://tinyurl.com/p6b6wpx.

Oldendick, R., A.L. Coker, D. Wieland, J.I. Raymond, J.C. Probst, B.J. Schell and C.H. Stopskopf, 2000. Population-based survey of complementary and alternative medicine usage, patient satisfaction, and physician involvement. CRVAW Faculty Journal Articles 135.

Ong, C.K. and B. Banks, 2003. Complementary and alternative medicine: the consumer perspective. The Prince of Wales's Foundation for Integrated Health, London, United Kingdom, 54 pp.

Osgood, C.E., 2009. Semantic differential technique in comparative study of cultures. American Anthropologist 66(3): 171-200.

Pietroni, P.C., 1992. Beyond the boundaries: relationship between general practice and complementary medicine. British Medical Journal 305: 564-566.

Rose, M., 1977. The horsemaster's notebook: revised edition. Harrap, London, United Kingdom, 230 pp.

Russell, L., 2003. Showing commentary. Horse and Hound, May, IPC Magazines, London, United Kingdom.

Roberts, M., 2002. From my hands to yours. Pat and Monty Roberts Inc., Solvang, CA, USA, 230 pp.

Shields, R., 1996. The individual consumptions cultures and the fate of the community. In: Shields, R. (ed.) Lifestyle shopping: the subject of consumption. Routledge, London, United Kingdom, pp. 99-113.

Trusov, M., R.E. Bucklin and K. Pauwels, 2009. Effects of word-of-mouth versus traditional marketing: findings from an internet social networking site. Journal of Marketing 73(5): 90-102.

Wallendorf, M. and M. Brucks, 1993. Introspection in consumer research: implementation and implications. Journal of Consumer Research 20(3): 339-359.

Part 6.
Governance issues

The next part focuses on *governance* and in particular, the relationship between the horse world and wider governments. Thus Crossman and Walsh compare that relationship in the UK, Sweden and the Netherlands. They speculate on issues which may arise from the need to make what have been traditional systems of governance in the horse world able to respond to the challenges of the 21st century. Jez *et al.* report from a project which asked both the public, and horse professionals to speculate on where the industry will be in three decades. Their reports provide plenty of food for thought, particularly as the sector has not been good at coming together to make decisive changes. If such actions do not occur, however, there is a risk that the equine sector will stumble blindly into situations not anticipated, creating difficulties and potential alienation from the greater society in which it is situated.

13. The relationship between the government and the horse industry: a comparison of England, Sweden and the Netherlands

G.K. Crossman[1]* and R.E. Walsh[2]

[1]equiRED, P.O. Box 293, Barnstaple, Devon, EX32 2BQ, United Kingdom; gkcrossman@gkcrossman.co.uk
[2]Royal Agricultural University, Stroud Road, Cirencester, Gloucestershire, GL7 6JS, United Kingdom

Abstract

In England, Sweden and the Netherlands, the relationships between the government and the horse industry have evolved over time. At the beginning of the twentieth century the horse was considered a beast of burden, with its role closely linked to agriculture, defence and transport. Just over one hundred years later its function has been transformed, with horses now predominantly involved in recreational and sporting activities. The government's view of and connection with the industry in each country that this research focuses upon reflects this change. Policy network theory was used to analyse the development of the relationship between the government and interest groups in each horse sector examined, with evidence gathered through interviews, documentary research and participant observation providing the basis for the study. A specific organisation which acted as the mouthpiece for the industry to the government in each country was identified. However, the structure of these bodies and their mode of operation differed considerably between countries. The level of government intervention and financial support afforded to each horse industry also varied. Links between existing policy areas and the horse industry were found to be significant in the success of the industry. In addition, in Sweden and the Netherlands the connection between the equine and agricultural sectors was important in the development of the relationship between the government and the horse industry.

Keywords: horse sector, equine policy network, Horse Council

Introduction

Once a small browsing mammal, only 22 centimetres tall and referred to as the 'Dawn Horse', the horse is now an animal with the ability to travel great distances and inherent strength. As a result of its evolution, it has formed a close relationship with man. Primarily considered a beast of burden, in Western Europe during the early 1900s the horse was a multi-functional, utilitarian animal providing a means to work the land, transport for both people and goods, and assistance in meeting defence and military requirements. By the end of the 1900s, the role of the horse in this region had dramatically altered, and it can now be found principally engaged in a variety of sporting and recreational uses (Crossman and Walsh, 2011). These changes have reconstructed the way in which the horse is viewed by society and government. Moreover, the organisational landscape of the horse sector in each of the three countries examined has expanded in recent years, due to this transformation in the function of the horse.

The aim of this chapter is to explore the relationship between the horse industry and the government in England, Sweden and the Netherlands, using policy network analysis as a contextual framework.

Research methodology

The initial research this chapter is based upon was gathered at the Centre for Rural Policy Research, Department of Politics, University of Exeter, from 2006 to 2010, through the first author's PhD 'The organisational landscape of the English horse industry: a contrast with Sweden and the Netherlands' (Crossman, 2010). The methodological approach underpinning the study is an analysis, through a case study, of the equine policy networks of the three selected countries, including an examination of their structure, component organisations and associated roles. The case study method of research focuses upon one, or sometimes a few, instances of a particular phenomenon in order to provide a detailed account of events, relationships, experiences or processes which occur during that particular instance (Denscome, 2007). This research focused upon the relationship between interest groups and the government in the horse industry of each country, and how this affects the policy process.

England, Sweden and the Netherlands were chosen as the countries to be included within the study. England was selected as the home country. Sweden and the Netherlands were picked from an initial long list of nine countries, six within Europe and three from outside the area. The selection process took account of a number of factors, including: the relevance of the country to the PhD stakeholders; availability of information within the country, for example how much text was available in English; whether the country contained features that, following analysis, might benefit the horse industry in England and its relationship with the government; the relevance of the country to possible organisations who might provide funding for the field trips; and whether these countries would enable the case studies to be completed in the time available.

Both primary and secondary data can be found within each case study. Primary data were gathered through 40 formal, semi-structured interviews, the majority of which were carried out face-to-face. Alongside these, informal discussions also took place. Participant observation, through a number of contexts including observation at conferences and other key equine policy network gatherings, provided a further source of primary data. Secondary data came from documentary analysis, including the examination of country-specific statistical and economic reports, journal papers and other appropriate literature. Policy documents, where available, were also evaluated. These documents had either been identified during the literature review, or were highlighted during the interviews or attendance at key equine policy network gatherings and were then sourced.

As well as an in-depth analysis of the horse industry in the home country of England, field trips to Sweden and the Netherlands were completed in order to gather as much rich and varied data as possible. A total of two and half months were spent in Sweden, through three trips taken in 2008 and 2009, including a visit to the EU Equus 2009 Conference held in Uppsala. Just over two weeks was spent in the Netherlands, in 2008.

Since the initial study the data gathered has been updated; findings are included within this chapter.

Policy networks

The contextual framework used within this chapter to analyse the relationship between the government and the horse industry within the selected countries is policy network theory. Developed during the 1970s and 1980s, policy network theory allows relationships, links and interdependencies between government departments and actors in a specified sector or area to be explored. Actors can be organisations, interest groups, or individuals. The theory examines the continuity within these relationships, highlighting how actors within these networks communicate, identify issues, take collective action and share resources (Marsh, 1998; Rhodes, 1997, 2006). Rhodes (2006, p.426) offers the following definition:

> Policy networks are sets of formal institutional and informal linkages between governmental and other actors structured around shared if endlessly negotiated beliefs and interest in public policy making and implementation. These actors are interdependent and policy emerges from the interactions between them.

Policy networks have been used to analyse a number of sectors, including agriculture, environmental planning and tourism (see, for example Burstein, 1991; Dredge, 2006; Duagbjerg, 1998; Selman, 2000; Smith, 1992).

Most policy networks contain a core and a periphery (Laumann and Knoke, 1987). Smith (1992) develops this concept to suggest policy networks have layers, where the primary layer comprises organisations and key actors who are significant in guiding policy decisions made by the government which impact upon the sector. These organisations and key actors are involved with each other on a day-to-day basis, setting the rules of the game and determining the membership of the network. The secondary layer consists of organisations and key actors who do not have a continuous influence on policy and are only active when certain issues in which they have an interest, or which present an issue for them, are discussed. Those in the secondary layer abide by the rules of the game, set by those in the primary layer, and often have limited resources.

Smith (1992) further suggests each policy network contains a number of 'elements' or 'sub-sectors'. While organisations and actors in the primary layer of each policy network are involved in all policy decisions, those in the secondary layer only contribute to the element or elements of the policy process that interest them.

The framework provided by policy network theory will now be used to conceptualise the horse industry, and the relationship between the government and interest groups within the sector, in the countries studied in this research.

The structure of the equine policy network

As described above, Smith (1992) highlighted the core and periphery, or primary and secondary layers, of a policy network. In order to represent the equine policy network it is necessary to expand this idea to include three principal layers: the primary layer, the secondary layer and the tertiary layer, along with the periphery which surrounds the network.

Within each equine policy network the primary, or core, layer contains the Horse Council of each country: the organisation that is formally recognised, by both the government and the industry, as providing the link between the government and the sector to the wider policy network. The secondary layer comprises a number of supporting organisations, which actively support the Horse Council through the provision of resources, in the form of funding, personnel or other means. The interest groups found in the secondary layer are powerful within the network and can significantly influence policy decisions. All other organisations involved in the policy making process can be found in the tertiary layer. These interest groups do not have the same level of power as those in the secondary layer, and therefore their influence on the policy making process is often reduced. The periphery contains those outside the policy network, who do not actively engage with it, and have no influence on policy making. However, depending on the issue at hand, actors in the periphery could have an influence over decisions made inside the network.

Smith (1992) also highlights a number of elements or sub-sectors within the policy network, suggesting that organisations in the secondary, and in this case the tertiary, layers of the network, contribute to the elements that interest them. Within each country studied a number of elements can be found in the equine policy network. These include: racing; sport and recreation; breeding; research; and education. It should be noted that this list is not exhaustive, and not all elements within each policy network hold equal importance across each country. For example, racing in the Netherlands is a very small part of the equine industry, and therefore is not as significant as it is in England or Sweden.

Examining the layers and elements of the equine policy networks as described above, and within this chapter, makes it possible to conceptualise their structure, as seen below.

At the centre of the diagram is the Horse Council, while supporting organisations are found in the secondary layer. Other interest groups inhabit the tertiary layer. Five different elements, or sub-sectors, are shown as an example in Figure 1. Actors might be interested in more than one element, and will move between them when necessary.

The mouthpiece of the horse industry

Within each country there is an identified organisation that acts as a mouthpiece for the horse industry and represents its interests to the government; these organisations are referred to as the Horse Council. The Horse Council often has a clear link to at least one government department, and will act as an intermediary, putting the government in touch with another actor within the network, if it cannot provide detail about a particular request or query.

In Sweden, the Swedish Horse Council (HNS) is at the core of the network. The British Horse Industry Confederation (BHIC), has been at the helm in recent years in England, although the Equine Sector Council (ESC) is now working alongside the BHIC to provide a mouthpiece for the industry to the government. While in the Netherlands, the Dutch Horse Council (SRP) holds the position. Each Horse Council will now be taken individually, and its evolution and mode of operation will be examined.

Sweden

The most mature Horse Council of the countries focused upon in this study is HNS. In Sweden, the rising popularity of the horse and its significance to the rural economy prompted a state inquiry by the Swedish Board of Agriculture (SJV) in 1990 and 1991. This identified the need to create a

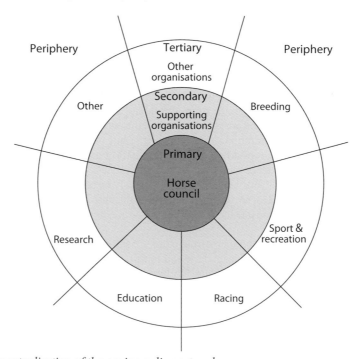

Figure 1. Conceptualisation of the equine policy network.

common platform for the horse and surrounding industry, thus beginning the formation of Sweden's current equine policy network (HNS, 2007; Karlander, 2002).

During 1992, a collaboration between the National Federation of Swedish Farmers (LRF) and the Swedish Horseracing Totalisator Board (ATG), together with support from organisations within the wider horse industry, resulted in the formation of the Swedish Horse Council (HNS). The structure of the Board of HNS allows interested organisations to have input into its membership.

From 1992 to 1995, HNS was staffed on a part-time basis. However, in 1995 this changed when a General Manager, Mr Olof Karlander, was employed. As time has passed HNS has grown, and now employs a number of full time staff, including Mr Stefan Johanson who took over from Mr Karlander as CEO in the late 2000's. When Mr Karlander was first employed he was given two priorities: to reorganise and modernise the National Estates (Strömsholm, Flyinge and Wången) and to implement the Horse Parliament.

The first Horse Parliament took place on 6th March 1996, when HNS went directly into the Swedish Parliament with the sole purpose of raising the profile of the horse industry. Prior to this meeting HNS had carried out an investigation into the size and scope of the horse industry, identifying its socio-economic importance to Sweden. Armed with this information, the government was urged to take a more active interest in the industry when it came to issues affecting the sector, such as agricultural policy. The Horse Parliament and the HNS investigation were both key in the advancement of the equine policy network in Sweden and in raising the profile of the horse industry with the government.

Following the first Horse Parliament, HNS have continued to build the awareness of the socio-economic benefits of the industry to the government, at national, regional and local level, and to other interested parties, including the agricultural sector and media. Further events similar to the first Horse Parliament have taken place, and HNS is actively engaged with the government both when there are issues of interest to the horse industry and in a wider context.

As a result of work undertaken to establish the socio-economic importance of the sector, HNS suggested there was a need to create a new policy area within the Swedish government specifically for the horse industry, alongside the existing 48 budgetary areas. However, this proved to be a challenge, and ultimately it was not possible to include a 49th area. After reconsidering its ambition, HNS adopted a different approach, studying the industry from a political angle in relation to the existing 48 areas, to consider where the horse industry aligned with them. The results showed the horse was significant to 12 areas, including rural development and education. HNS then contextualised the equine industry in these areas to make it more accessible to the government.

The majority of HNS' funding comes from ATG. In 2013, HNS was provided with a total of SEK 50 million (£4.68 mil / €5.61 mil, OANDA, 2014) from ATG, through the contract held between the Swedish State and the Swedish Trotting Central Association (STC) and Swedish Thoroughbred Racing (SG) (HNS, 2014). This contribution from gambling is spread across four areas, as follows:
- educational programmes: SEK 33 million (£3.09 mil / €3.70 mil, OANDA, 2014);
- youth programmes: SEK 6 million (£0.56 mil / €0.67 mil, OANDA, 2014);
- development projects, including the evolution of the industry, and breeding and production of Swedish horses: SEK 6 million (£0.56 mil / €0.67 mil, OANDA, 2014); and
- national estates, including Flyinge, Strömsholm and Wången: SEK 5 million (£0.47 mil / €0.56 mil, OANDA, 2014)

HNS is also actively involved in the wider equine policy network, as a founding member of the European Horse Network (EHN): a not-for-profit network of stakeholders acting at a world, European, national or regional level within the European horse sector (EHN, 2014).

England

The Horse Council in England, which covers Britain, is a few years younger than HNS. In 1997, Lord Bernard Donoughue of Ashton, then Minister for Farming and Food, tried to get the Ministry for Agriculture, Food and Fisheries to incorporate a Ministry for the Horse, and also a Minister to take responsibility for the Horse. He himself succeeded in becoming the first Minister for the Horse, but no Ministry was created (Donoughe, 2003).

Following this, in 1999, the British Horse Industry Confederation (BHIC) was formed, when Lord Donoughue suggested various interested parties form an umbrella organisation to fully represent their diverse interests. This was the first time the racing and non-racing sub-sectors had been brought together in England, through the involvement of the British Horseracing Board (BHB, now known as the British Horseracing Authority, BHA) and the British Equestrian Federation (BEF) (Donoughe, 1999).

Upon its creation, the BHIC included the BHB, BEF and Thoroughbred Breeders' Association (TBA) in full membership, while the British Horse Society (BHS) was formally acknowledged within the body. These four organisations each appoint one Director to the Board of the BHIC: the BHA and TBA each have a Director, while the BEF has two Directors, one of whom is appointed from the BHS. After its initial formation, the BHIC identified that it did not represent the whole industry, and the British Equestrian Trade Association (BETA) and British Equine Veterinary Association (BEVA) were appointed as additional member organisations, without Director rights, in order to increase representation from interests within the sector.

Unlike HNS in Sweden, the BHIC has no direct funding, rather it relies on resources from its six component organisations. These organisations financially support representatives to attend meetings, and the BEF provides the services of its Head of Finance as Company Secretary. It is led by a nominated Chair, on a rolling (normally) two-year tenure. The post rotates between the racing (BHA and TBA) and non-racing (BEF and BHS) sub-sectors.

In 2005, the Strategy for the Horse Industry in England and Wales was published by the BHIC, along with Defra (Department for the Environment, Food and Rural Affairs), DCMS (Department for Culture, Media and Sport) and the Welsh Assembly Government. This document reinforced the role of the BHIC not only as the mouthpiece of the horse industry to the government, but also in pulling the sector together to represent their interests to the government. However, the success of this document, which contained a ten year strategic plan based around eight aims, has been somewhat hit and miss. Where specific aims were written alongside the strategy of the organisation, or part of the organisation, which had responsibility for it, the level of achievement can be shown to be greater than where there is less of a relationship (for example, many of the action points included in Aim 8 'Improve the quality and breeding of horses and ponies' have been achieved as they run in parallel with the strategy for the Equine Development portfolio of the BEF).

During May 2010 a general election saw a change in government in the United Kingdom, from Labour to a Conservative-Liberal Democrat coalition, and the role of the Minister for the Horse ceased to exist. This appointment had continued through a number of incumbents and guises from Lord Donoughue in 1997, always including responsibility for the horse industry, and at times encompassing authority for equine health and welfare. At the time the role ceased, the areas of the horse industry and equine health and welfare were joined together under the umbrella of the Minister for the Horse Industry.

Following the inevitable changes in policy and direction brought about by the new regime in the United Kingdom, the relationship between the government and the horse industry slightly altered. In

April 2011 the Animal Health and Welfare Board for England (AHWBE) was formed. The creation of this Board came about as a result of the recommendations of the Independent Responsibility and Cost Sharing Advisory Group published in December 2010, which gave clear guidance to Defra Ministers on strategic policy affecting the health and welfare of kept animals in England. Horses were included in these recommendations, alongside farm animals and pets (Defra, 2012, 2014).

In order to work effectively with the new AHWBE, the Equine Health and Welfare Strategy (EHWS) Group became the Sector Council for Equine Health and Welfare (Equine Sector Council or ESC) in 2012. The Sector Council concept is one adopted by Defra across all kept animal sectors. The EHWS organically fulfilled the position for horses, in relation to equine health and welfare. Other examples include the Companion Animal Sector Council (CASC) and the Pig Health and Welfare Council (PHWC).

The ESC has a direct link to Defra through the AHWBE, and deals with them in relation to equine health and welfare. Working closely alongside the ESC, the BHIC continues to provide the mouthpiece for the industry to the government in matters of policy development and implementation relating to equestrian sport and recreation, breeding and trade (BHIC, 2012).

The Netherlands

The youngest Horse Council of the countries featured in this study is SRP, which developed in two distinct stages. Its initial incarnation was formed through a collaboration of interests across the horse industry, when it was established as a member for the Product Boards for Livestock, Meat and Eggs (PVE) (Van Lenthe and Van Markus, 2000). However, the Dutch government did not formally recognise this organisation as representing the horse industry, rather it considered the horse industry as disorganised, encompassing a large number of people with their own interests.

The second stage of SRP's development was the result of the establishment in 2003 of the Federation of Dutch Horse Entrepreneurs (FNHO), along with, in 2005 and 2006, the increased interest of the Dutch Federation of Agricultural and Horticultural Organisations (LTO) in the horse industry and the recognition by the government that horses were becoming more socially and economically important. No longer linked to PVE, on 1st January 2007, SRP was formally recognised by the government and the horse industry as an independent, non-commercial organisation.

In 2007, the Board of SRP contained representation from key organisations and interests within the Dutch horse industry. FNHO represented the 'entrepreneurial sector' of the horse industry, including the Federation for Dutch Riding Sport Centres (FNRS), Instructor, Sport and Training Stables (IST), the Federation of Dutch KWPN Stallion Owners (FBvHH), the Association of Studs and Breeders of Horses (VHO) and the Association of Horse Dealers Netherlands (VSN). The Royal Warmblood Studbook of the Netherlands (KWPN), along with one representative from the horse studbooks and a further representative from the pony studbooks, spoke for the breeding sub-sector. Sport was represented by the Royal Dutch Equestrian Federation (KNHS) and Dutch Trotting and Thoroughbred Racing (NDR). The Board of SRP also contained representation from LTO. Since its establishment in 2007, there have been some changes in these organisations as they have developed: for example the interests represented by FNHO have expanded. However, the principal of the coverage of the horse industry within SRP remains the same.

LTO has played a significant role in the development of SRP. When SRP was re-established in 2007 it was funded by LTO, with its Secretary seconded from LTO. This arrangement was in place for an initial two years, with the funding after this time coming from SRP's constituent organisations, which still included LTO. SRP's Secretary has an administrative role, along with responsibility for specific aspects of SRP's portfolio, for example as a member of the Working Group on Social,

Economic and Fiscal Policy, while members of SRP's Board and supporting organisations also have input into the range of working groups and actively engage with representatives of the Dutch central government and other interested parties.

Important factors for the equine policy network

Through the development of the equine policy networks studied within this research it is possible to identify important factors that should be considered. These factors include: the effectiveness of the Horse Council; policy alignment; the relationship between the equine and agricultural policy networks; and communication. Each will now be briefly discussed.

The effectiveness of the Horse Council

The role of the Horse Council within the equine policy network is two-fold. Firstly, it is the mouthpiece of the horse industry, both to the government and for other interested parties when relevant. Secondly, it has a pivotal position in coordinating the interests of all those within the equine policy network. When both of these functions are carried out successfully the benefits to the horse industry are apparent.

The pivotal contribution HNS provided in building awareness of the socio-economic benefits of the sector to the government, at national, regional and local level, and to other interested parties in Sweden, presents an excellent example of the importance of the role of the Horse Council. Within the equine policy network in England, the BHIC effectively united the sector in the production of the Strategy for the Horse Industry, another example of the importance of the Horse Council. The second incarnation of SRP also had a very positive effect on the equine policy network in the Netherlands. The government changed their perception of the sector from believing it was disorganised, encompassing a large number of people with their own interests, to recognising the socio-economic contribution of the industry and giving it a legitimacy and level of coordination unseen before.

In order to be effective and successfully carry out these functions, the Horse Council needs to be properly resourced, and able to work effectively with all those in the equine policy network.

Policy alignment

Policy alignment within the equine policy network, and in a broader governmental context, has been found to be invaluable within this study. In Sweden HNS' contextualisation of the equine sector, in relation to the 12 governmental budgetary areas where the horse is considered to be significant, has been advantageous to the industry. As a result, the socio-economic benefits of the horse industry have been more easily understood, both by those within government and outside of it, and the sector has been able to reap the rewards of this.

Evidence of the benefits of policy alignment can also be seen through the Strategy for the Horse Industry in England and Wales. This document has made the most progress where it has been aligned with the strategy and policy of the organisation which has been assigned responsibility for each action point.

The relationship between the equine and agricultural policy networks

In the past 20 years or more the horse industry has played a considerable role in farm diversification across Western Europe, as many agricultural holdings have branched out to incorporate some form of equine enterprise on their land. The significance of the horse industry to agriculture has been identified in all three countries studied in this research (see, for example, Crossman and Walsh, 2008; Häggblom *et al.*, 2005; Turner *et al.*, 2003).

Within Sweden and the Netherlands the involvement of the key agricultural organisation (LRF in Sweden and LTO in the Netherlands), already in established relationships with their respective governments through their agricultural policy networks, expedited the development of the relationship between the government and the horse industry in the equine policy network. Both LRF and LTO had recognised the growing, significant role of the horse to their agricultural sectors. They identified that assistance provided to the horse industry would also benefit their primary focus, the agricultural sector.

The actions of LRF and LTO, in assisting in the development of the Horse Council, have resulted in a strong legacy for these organisations within the horse industry. Both have a clear, direct and established link to the Horse Council, which will reinforce their relationship within the equine policy network. When issues arise that affect the equine policy network, in which they have an interest, LRF and LTO both have direct access to the Horse Council to get their views across.

Conversely, in England there is no direct link between the equine and agricultural policy networks. The horse industry has been shown to be equally as important to agriculture, through the development of equine enterprises alongside existing agricultural businesses, as in Sweden and the Netherlands. However, the lack of a direct link between these two sectors makes it difficult for them to work together easily for their mutual benefit.

Communication

Each equine policy network comprises a number of stakeholders, with varying interests and motivations, along with differing levels of involvement, as indicated by the primary, secondary and tertiary layers. The periphery contains those outside of the network, who do not actively engage with it.

It has been identified that the horse industry makes a substantial socio-economic contribution to the countries included in this research; therefore it is imperative that each equine policy network takes steps to safeguard its sector. A large proportion of those to whom policy decisions relate can be found in the periphery and it is important that their interests and motivations are considered. This is particularly significant when considering policy and issues in relation to equine health and welfare, as this can be found across all elements and sub-sectors of the network. In order to ensure that as many of those as possible are aware of any policy that could impact upon them information should be clearly communicated, in a language and way that is accessible and appropriate to them.

A distinction in communication should also be made when dealing with the equine or agricultural sectors. Whereas the majority of farmers will own many head, for example of cattle or sheep, it has been suggested that in Britain approximately 65% of horse owners own one equine (BHIC, 2009). It is likely that this finding would be mirrored in Sweden and the Netherlands if similar data were available. The motivations and commitments of people keeping horses, compared to those who own agricultural animals, are also fundamentally different.

Where the equine and agricultural policy networks are closely aligned, work well together, and the government is clear about the requirements of each, it is likely to be easier to target a message to those for whom it is intended. However, if the two policy networks do not have a clear relationship, and the government is unsure about the requirements of each, communication will be harder and this will have a negative impact on the sectors.

Conclusions

In conclusion, each of the countries studied has identified the significant socio-economic contribution made by the equine sector, and this needs to be built upon. The relationship between actors within the equine sector and the government is of great importance. The role of the Horse Council in the success of this relationship is pivotal, both in being the mouthpiece for the industry, and in coordinating and directing communication between the government and relevant organisation(s) within the policy network when appropriate.

The linkage between the equine and agricultural policy networks is also significant, as there is a level of interdependence between the two sectors. Where a formal relationship has been established, both sectors have benefitted. However, where no formal relationship exists it is harder for the two sectors to work in collaboration, and this could be viewed as detrimental to both.

Policy alignment, whether between the equine sector and government, or within the horse industry, has been shown to be advantageous for the equine policy network. Where possible, this should be built upon. Effective communication, at all levels, which takes into account the interests and motivations of the varied stakeholders, is also vital. This should be considered within the policy network, and with those on the periphery, in order to ensure that any messages have as far-reaching an impact as possible.

Acknowledgements

Some of the content of this chapter is taken from Georgina Crossman's PhD thesis 'The organisational landscape of the English horse industry: a contrast with Sweden and the Netherlands,' which was completed at the Centre for Rural Policy Research, Department of Politics, University of Exeter, United Kingdom. The authors would like to acknowledge the role of Prof Michael Winter and Dr Matt Lobley in the writing of the PhD. The PhD studentship was jointly funded by the Department for Environment, Food and Rural Affairs (Defra), British Equestrian Federation (BEF), the Glanely Trust at the University of Exeter and the Royal Agricultural College (now the Royal Agricultural University). Additional funding was provided by the Stapledon Memorial Trust, the British Society of Animal Science (Murray Black Award) and the BEF for field trips to Sweden and the Netherlands. The Department of Economics at the Sweden University of Agricultural Sciences, Ultuna, and Van Hall Larenstein University of Applied Sciences, Wageningen, are also acknowledged for their role in assisting in the gathering of data.

References

BHIC, 2009. BHIC briefing – size and scope of the equine sector. BHIC, London, United Kingdom.

BHIC, 2012. News bulletin – January. BHIC, London, United Kingdom.

BHIC, Defra, DCMS and the Welsh Assembly Government, 2005. Strategy for the horse industry in England and Wales. Defra, London, United Kingdom. Available at: http://tinyurl.com/qxjst4v.

Burstein, P., 1991. Policy domains: organisation, culture, and policy outcomes. Annual Review of Sociology 17(1): 327-350.

Crossman, G.K. and R.E. Walsh, 2008. The role of the horse industry in farm diversification. In: Proceedings of Roots 2008 – Conference of the Royal Institute of Chartered Surveyors. RICS, London, United Kingdom.

Crossman, G.K. and R.E. Walsh, 2011. The changing role of the horse: from beast of burden to partner in sport and recreation. The International Journal of Sport and Society 2(2): 95-110.

Crossman, G.K., 2010. The organizational landscape of the English horse industry: a contrast with Sweden and the Netherlands. University of Exeter, Exeter, United Kingdom.

Daugbjerg, C., 1998. Similar problems, different policies: policy networks and environmental policy in Danish and Swedish agriculture. In: Marsh, D. (ed.) Comparing policy networks: public policy and management. Open University Press, Buckingham, United Kingdom, pp. 75-89.

Defra, 2012. Animal Health and Welfare Board for England: about us. Available at: http://www.defra.gov.uk/ahwbe/about/.

Defra, 2014. Animal Health and Welfare Board for England. Available at: www.defra.gov.uk/ahwbe.

Denscombe, M., 2007. The good research guide: for small-scale social research projects. Open University Press, Buckingham, United Kingdom, 392 pp.

Donoughue, B., 1999. Opening address, National Equine Forum. In: Suggett, R.H.G. (ed.) Proceedings of the 6th National Equine Forum. NEF Organising Committee, Wellesbourne, United Kingdom.

Donoughue, B., 2003. The heat of the kitchen: an autobiography. Politico's, London, United Kingdom, 392 pp.

Dredge, D., 2006. Networks, conflict and collaborative communities. Journal of Sustainable Tourism 14(6): 562-581.

EHN, 2014. European Horse Network. Available at: http://www.europeanhorsenetwork.eu.

Häggblom, M., L. Rantamäki-Lahtinen and H., Vihinen, 2005. Equine sector comparison between the Netherlands, Sweden and Finland. MTT Agrifood Research, Finland.

HNS, 2007. Verksamhetsplan: 2008-2010. HNS, Stockholm, Sweden.

HNS, 2014. The Swedish Horse Council Foundation (HNS). Available at: http://nshorse.se/in-english.

Karlander, O., 2002. The Swedish horse industry. In: Suggett, R.H.G. (ed.) Proceedings of the 10th National Equine Forum. NEF Organising Committee, Wellesbourne, United Kingdom.

Laumann, E.O. and D. Knoke, 1987. The organisational state. In: Smith, M.J. 1993. Pressure, power and policy: state autonomy and policy networks in Britain and the United States. University of Pittsburgh Press, Pittsburgh, PA, USA, 262 pp.

Marsh, D., 1998. The development of the policy network approach. In: Marsh, D. (ed.) Comparing policy networks: public policy and management. Open University Press, Buckingham, United Kingdom, pp. 3-17.

OANDA, 2014. Historical currency exchange rates. Available at: http://www.oanda.com/currency/historical-rates.

Rhodes, R.A.W., 1997. Understanding governance: policy networks, governance, reflexivity and accountability. Open University Press, Buckingham, United Kingdom, 252 pp.

Rhodes, R.A.W., 2006. Policy network analysis. In: Moran, M., M. Rein and R.E. Goodin (eds.) The Oxford handbook of public policy. Oxford University Press, Oxford, United Kingdom, pp. 427-447.

Selman, P., 2000. Networks of knowledge and influence: connecting 'the planners' and 'the planned'. Town Planning Review 71(1): 109-121.

Smith, M., 1992. The agricultural policy community: maintaining a closed relationship. In: Rhodes, R.A.W. and D. Marsh (eds.) Policy networks in British government. Clarendon Press, Oxford, United Kingdom, pp. 27-50.

Turner, M., D.M. Winter, D. Barr, A. Errington, M. Fogerty, M. Lobley and M. Reed, 2003. Farm diversification activities: benchmarking study 2002, final report to Defra. University of Exeter, Exeter, United Kingdom.

Van Lenthe, A.H. and R.C. Van Markus, 2000. The social and economic significance of horse production in the Netherlands. In: Book of abstracts of the 51st Annual Meeting of the EAAP. Wageningen Academic Publishers, Wageningen, the Netherlands, p. 367.

14. The French horse industry in 2030: scenarios to inform decision-making

C. Jez[1]*, B. Coudurier[2], M. Cressent[3], F. Mea[4], P. Perrier-Cornet[5] and E. Rossier[4]

[1]INRA, Department of Scientific Expertise, Foresight and Advanced Studies, 147 rue de l'Université, 75338 Paris cedex 07, France; christine.jez@jouy.inra.fr
[2]INRA, Management Board Unit (CODIR), 147 rue de l'Université, 75338 Paris cedex 07, France
[3]IFCE, Research and Innovation Department, 61310 Exmes, France
[4]IFCE, 83-85, Bd. Vincent Auriol, 75013 Paris, France
[5]INRA, Joint Research Unit for Markets, Organisations, Institutions and Actor Strategies (UMR MOISA), 2 Place Viala, 34060 Montpellier Cedex 2, France

Abstract

In France, the horse population has been increasing since 1995, and reached 950,000 horses in 2010. This represents about 15% of the current total horse population in Europe. This growth is the result of the development of pony-riding for children and the increasing interest of French people in recreational riding and in horserace betting. These changes create new opportunities for various sections of the horse industry, especially in the context of declining government support, increasing international competition in the horse market, social changes with regard to animal welfare, decreasing horsemeat consumption, and the harmonisation of regulations across the European Union. To understand new possible directions for research and for public policy, and also to help horse industry stakeholders prepare for future change, the French National Institute for Agricultural Research (INRA), in cooperation with the French Institute for Horse and Riding (IFCE), conducted a joint scenario-building exercise for the French horse industry to 2030. The study's basis was a 'morphological analysis', a method that explores past and current trends and potential shifts to consider possible future developments. It led to four strongly contrasting long-term scenarios ('Everyone on horseback', 'The high-society horse', 'The civic horse', and 'The companion horse'), which differed in terms of horses' uses, horse populations, and employment opportunities. However, all scenarios raised common concerns with regard to (1) the relationship between humans and horses; (2) economic efficiency; (3) environmental issues; (4) the preservation of horse breeds; (5) land-use pressures; and (6) the health, welfare, and care of animals up to and beyond death. These questions call for new research developments in the fields of animal behaviour, economic and social sciences, breeding, and genetic improvement. They also emphasize the need to enhance knowledge and innovation transfer.

Keywords: horse, horse industry, foresight, future, France

Introduction

Horseriding in France has long been unnecessary, yet horseriding has become an increasingly popular leisure activity while the horserace industry benefits from the wide appeal of horserace betting. Nevertheless, increasing competition from other countries, declining government support, changes to EU tax policy and to regulations governing the introduction of competition from online betting all cast doubt on the future of the industry and on the persistence of current trends. Beyond these concerns, the French horse industry must also confront a number of other issues related to employment, profitability, breed diversity, and animal welfare.

To improve the services they provide and to plan for future needs in knowledge and innovation, the French National Institute for Agricultural Research (INRA) and the French Institute for Horse and Riding (IFCE) agreed to jointly study the medium-term outlook for the French horse industry. This article begins by reviewing the methodology used and provides an overview of the industry at present before putting forward the various key elements seen to be influencing the future of the industry and the foresight scenarios. A comparative analysis of these elements follows and is presented in terms of lessons and opportunities for research and development.

Methodology: exploring possible futures through scenario-building

A foresight study is a process that investigates possible futures to inform action in the present. The exercise neither claims to provide quantified or dated representations of a future situation, nor intends to scientifically dictate future options to decision makers. It is an exploratory approach designed to bring clarity to current action in light of possible futures. The scenario-building process is based on identifying and investigating significant trends and areas of substantial uncertainty, risk, or possible discontinuity. The process draws on a method known as 'morphological analysis' (Börjeson *et al.*, 2006; Godet, 2000) that allows possible futures to be investigated by combining the development hypotheses of key elements affecting the future of the system being studied (De Jouvenel, 2000). From the many possible combinations, the choice of disparate exploratory scenarios is based on their relevance for action, their coherence, their plausibility (regardless their probability) and their contrast, so that they can offer a large vision of possible futures, sometimes by being closed to caricature. As a result, scenarios describe stories, eliciting new and original visions and suggesting paths of strategic analysis. They allow for relevant stakeholders and decision makers to be alerted to emerging phenomena and issues, to anticipate their consequences, and to devise proactive strategies (De Jouvenel, 2000).

The study was prepared by a project team and a working group made up of experts from different backgrounds and disciplines. To further strengthen the diversity of participants and of opinions, the team's work was broadly supplemented by interviews with economic and institutional stakeholders and by workshops held in three French regions. France was used as the geographical scale for the study's scenarios, and the country's diversity was represented using regional analyses. This scale bears witness to the wide range of scenario outcomes, notably in terms of the industry's distribution in urban, peri-urban, or rural areas, its value to tourism, the place of equestrian activities in local culture, and on land development pressures. The 2030 timeframe was chosen because it allows for investigation of existing trends and of early signs of change with potential to develop into causes of discontinuity, while also remaining a tenable time period for decision-makers.

The French horse industry in numbers

At the end of 2010, the population of horses in France was estimated at 950,000 horses for more than 1.5 million riders (Réseau REFErences, 2011). Of the 1.5 million riders in 2011, more than 700,000 held riding licences (FFE, 2011). The horse industry as a whole brings together approximately 53,000 businesses (Heydemann *et al.*, 2011). There is considerable diversity in these businesses, 85% of which are directly involved in the production, sale, or use of horses. For the most part, they are small, often family-owned, businesses. They employ around 72,000 people, most of whom are young and are increasingly women. Turnover in these business tends to be high (Heydemann *et al.*, 2011). While breeders represent 64% of horse industry business, the majority of jobs are found in horseracing stables and equestrian centres. Only 12% of total industry jobs are on breed farms (Heydemann *et al.*, 2011; Réseau REFErences, 2011).

Through its commercial activities, revenues for the horse industry in France exceeded four billion euro in 2011. This figure excludes revenues from horserace betting, which itself generated just over ten billion euros, seven billion of which was paid out to bettors as winnings (REFErences, 2012).

The future of the horse industry in France: five key areas of uncertainty

In examining the current state of the industry, its past development trends, and ongoing debates as to its future, approximately 40 elements likely to have an impact on the future of the industry were identified and grouped into five categories. These served as the bases for the scenario-building exercise.

Economic and social context: What conditions will the industry see over the next 20 years?

The history of horses, and the use made of them, has been strongly influenced by wider economic and social developments, including the motorisation of agriculture and transport (Rossier *et al.*, 1984), the changing place of leisure in society, women's liberation, and the increase of middle-class purchasing power. In the future, powerful development pressure for arable land, caused by a growing world population, is likely to have an appreciable impact on animal industries (Paillard *et al.*, 2010). Accompanying these global changes is the very uncertain outlook for economic growth in France. A hypothesis meriting further investigation as to its potential impact on the use of horses is an ongoing economic downturn affecting middle classes more substantively and for a longer duration. Changing social expectations will also play a key role, notably with regard to the increase in free time and the proliferation of available activities with which to fill it (Viard, 2006), and to the evolving relationships between people and nature and with one another.

A population profoundly attached to horses, with a particular emphasis on leisure activities

Although they are no longer used as a source of physical force, a tool of war, or a means of transport (Digard, 2004), horses continue to hold popular appeal, even among people who are not horseriders. Horses have been a source of dreams and aspirations for thousands of years (Wagner, 2005) and it seems highly unlikely that horses will disappear from French people's imaginaire. But the ways people use horses may vary considerably with time as a function of the economic, political or environmental conditions and with the importance given to beneficial characteristics that are not fully exploited, notably the potential of horses in terms of education and building social bonds.

Recent trends in society and growing concern for environmental protection and sustainable development seem to be powerful forces driving the diversification in the use of horses for territorial development. However, the current prolonged economic crisis and reduced public funding could hinder such projects, which are not always cost-effective from a purely economic perspective.

By contrast, the horsemeat industry offers very little by way of development opportunities. Consumption of red meat has been on the wane for a number of years now and only a very small percentage of the French population eat horsemeat at present (FranceAgriMer, 2010).

People interested in betting constitute a world apart. The social appeal of horserace betting is not as certain, given the development of online gambling. In recent years, the popularity of horserace betting has soared through the existence enterprising and creative offers. However, its popularity remains contingent on the economic environment and on the habits and changing tastes of different age groups.

Policy in transition and animal welfare regulations strengthening

For 15 years, the French Government has endeavoured to improve the efficiency of its public spending and sought to be more in line with European Union regulations. This has led the government to review its role in, and administration of, the horse industry(Ministère des sports *et al.*, 2003). The government will, however, retain its human and animal health responsibilities, in connection with horses and their use.

Public policy will increasingly be confronted with changing social concerns with regard to animal welfare and the consequences thereof to regulations, particularly at EU level. Risk reduction and accident prevention for both horses and riders, and ownership and transport conditions, will doubtlessly remain key issues in this area, while the debate on the ultimate question of the horsemeat industry will continue, touching on transport, slaughter, the status of animals, and on codes of conduct dealing with horses (Jez *et al.*, 2013a).

Equestrianism and horseracing

The large-scale development of equestrian centres over past decades has coincided with profound changes in equestrianism, teaching methods, the centres themselves, and also in their patrons. The development of pony riding has attracted a new type of very young, female customer that has since become the primary patron of equestrian centres (Tourre-Malen, 2009). Such new opportunities for growth will require not only updated facilities and teaching methods, but also a reconsideration of fee structures.

Outside of national programmes and schemes, equestrianism is becoming more popular and more diverse, particularly thorough tourism, trail riding, and equestrian performances. This is because of increased leisure time in contemporary lifestyles (Boyer, 2012; Digard *et al.*, 2004; Vial *et al.*, 2011).

Changes in horseracing are primarily driven by changes in the betting world, although they are also influenced, more indirectly, by the ability to raise racehorses and, by extension, the existence of people willing to buy them. Overall balance in the horseracing industry is therefore susceptible to changes in economic conditions and betting volumes, but also to the ability of the system to distribute attractive amounts of prize money to the owners of winning horses.

In the absence of political will or financial support from local governments, the development of other horse-based activities, like equine therapy, mediation, rehabilitation, education, and use for municipal services, has been slow.

Horse breeding

Horse breeding and rearing in France has for centuries been carried out by a wide range of different people, most of whom were rarely specialized in the activity (Roche, 2008). Even today, 90% of broodmare owners breed horses as a hobby rather than as a source of income. Consequently, breeding is mostly driven by the pleasure derived from raising a foal and by the dream of producing a future champion (Couzy, 2007). Given the ongoing cuts to government support and the growth of competition, these practices are again giving rise to tension between breeders trying to earn a living from this activity and those for whom breeding is a hobby.

There is no general trend towards more economically efficient breed farms. Innovations to improve selection, breeding techniques, or animal assessment have not been equally received by all stakeholders (Jez *et al.*, 2013a).

Within 20 years, it will probably be possible to analyse the genetic profile of a horse to determine its potential for various activities. Will breeders in France opt to use such new techniques to determine at an early age the most suitable vocation for a horse? Given the current state of affairs, the answer is unclear. Will economic or regulatory pressures lead to critical thinking about technology and economic imperatives?

Four distinctive exploratory scenarios to 2030

Working from the major areas of uncertainty, four distinctive scenarios were developed. They put forward four possible futures for the French horse industry in 2030. The first two scenarios, 'Everyone on Horseback' and 'The High-Society Horse', are based on opposite developments in the economic climate. In 'Everyone on Horseback', the recovery of purchasing power facilitates access to leisure activities, whereas 'The High-Society Horse' sees ongoing economic downturn limit access to certain segments of the population. The two following scenarios, 'The Civic Horse' and 'The Companion Horse', are based on social changes, with a notable emphasis on quality of life and social solidarity in the former, and on growing concern for animal welfare in the latter (Jez *et al.*, 2013b).

Scenario 1 – Everyone on horseback. Horses in the leisure market, buoyed by a wide variety of businesses

After a period of gradual economic recovery, middle class purchasing power has increased; desires to ride horses at any age or skill level can finally be fulfilled. The government no longer provides direct support to the industry, which has the economic autonomy to meet the diverse expectations of its growing user population. Expansion of the equestrian recreation market has stimulated competition and encouraged diversity in equestrian activities to satisfy all riders, from the occasional tourist to the regular competitor. Furthermore, thanks to the success of marketing operations targeting young bettors, horseracing is flourishing and the PMU (*Pari Mutuel Urbain*) is Europe's leading horserace betting operator. The boom in equestrian activity benefits the industry as a whole, but requires it to become more professionally oriented and to make significant efforts to reduce costs while continuing to improving supply. In order to compete with the growing import of horses from abroad, new breeders have turned their production towards lower-cost horses bred for recreation rather than for competition. In reaction, producers of top-quality horses are spurred into making better use of horses not selected for competition by determining the most suitable vocation for each animal at an early age. This new organization of the industry is based on the connection between user expectations and horse breeding. Horses produced by the breeding industry in France are once again favoured by riders and are sought after on international markets.

Scenario 2 – The high-society horse. Limited number of users in a socially divided society where natural resources are under pressure

The lasting economic crisis has profoundly affected the middle class whose spending power has declined. In parallel, tensions over access to land and agricultural raw materials are driving up the cost of equestrian activities. The middle and working classes therefore choose more affordable recreational pursuits and riding is once again a social marker attracting a wealthy and exacting clientele. Well-known and innovative equestrian centres have adapted to meet the needs of these customers seeking the atmosphere of an elite club. In contrast, economic difficulties encourage gambling, including betting on horses. Horseracing remains financially stable through levies on bets placed, the use of modern technology, and growth in international betting. Facing rising structural costs that are already very high, racehorse trainers become themselves owners and even breeders. Harness racing nevertheless remains more robust than flat racing. The horsemeat industry disappears without any alternate use found for draught horses. Breeding becomes more focused. In order to remain viable, riding-horse breeders have been obliged to specialize and become more professionally oriented by producing elite sport horses selected at an earlier age on the basis of their potential. Other breeders have stopped breeding. With better control of breeding techniques and of economic considerations, and with suitable vocations found for horses at an early age, French horse breeders are able to build a place for themselves on international markets.

Scenario 3 – The civic horse. The horse in public and collective action, the link between humans, the land, and nature

In a climate of globalisation and an accelerated pace of life, there is a growing desire to reconnect with nature and with other people. Local and regional governments respond by backing initiatives promoting horse use to develop the economy, to protect the environment, and to contribute to individual and collective wellbeing. European Union grants, decentralised national funding, and the restitution of government racetrack betting levies (over which it again has a monopoly), enable contracts to be signed between municipalities and private and community-based equestrian businesses. Although riding for everyone is at the heart of these measures, also strongly promoted are the use of horses to maintain the landscape, for farming, for social and community services, for equine therapy, for education, and for social inclusion. The most popular and profitable racecourses are incorporated into an approach to develop France's riding and horse sport heritage, while the others are forced to close. This multiplicity of uses allows a horse to be used for a number of vocations over the course of its life, thereby reducing the overall need for animals. The French market is essentially supplied by qualified French breeders who sell products adapted to the new demand through local networks, developing quality symbols and discipline aptitude certifications.

Scenario 4 – The companion horse. From exploiting to caring for animals – the quest for animal welfare

Society's perception of nature and animals has evolved considerably over the last few decades. The horse has become a lifetime companion that people love to care for, and is respected in its own right. Owners visit equestrian centres to learn about horse behaviour and how to look after their horses, and are introduced to activities on foot or on horseback practised in competitions. In parallel, the population's concern for animal welfare has led to a ban on the consumption of horsemeat and regulations concerning equestrian practices have been tightened. As a result, owning and breeding are regulated while rules governing equestrian competitions and horse sports have been tightened with an aim to reduce animal suffering. The racing sector, already transformed by the complete liberalisation of betting, has gone further, organizing virtual races in parallel to real ones that take place on about twenty high-tech racecourses integrated into casinos. Breeding horses is above all a leisure pursuit for private owners. Nevertheless, some breeders who were previously passionate about performance now produce horses capable of long sporting careers and, at the same time, with specific aesthetic characteristics. A market for inactive horses has developed on the Internet, meeting the growing demand for horses as companions. The horse population is ageing and is replaced very slowly.

Contrasting perspectives of the four scenarios

Horse populations differ across the four scenarios; they may be double current levels in 'Everybody on Horseback' or half current levels in 'The High-Society Horse'. The scenarios also offer differing views of the types of horse that will be popular and what kind of characteristics they will have, although all agree that horses with good temperaments will be essential and that we should aim for a better early temperament analysis to more effectively orient horses towards their abilities. The businesses and types of employment involved in the industry also vary considerably, as does the distribution of horses across France's regions and rural areas. They are differences both in terms of the type of space i.e. rural, peri-urban and suburban, and the regions affected (Table 1).

Table 1. Impact of scenarios on the horse industry in 2030.

	Everyone on horseback	The high-society horse	The civic horse	The companion horse
Users	Clientele from the middle classes, all ages, men and women	Wealthy clientele, gender parity re-established	Varied beneficiaries from every social class, men and women	Varied, mainly female clientele without distinctions of social class
Horses	The horse is a medium for recreation · Numbers ↗↗↗ · Clear segmentation between recreation/sport/racing · Considerable import followed by domestic procurement	The horse is a social marker · Numbers ↘↘↘ · Specialization – top-quality sport and racehorses · Dynamic export of top-quality horses	The horse is a vehicle for social bonding · Numbers ↘ · Horses guided towards, reared, and adapted for specific and varied uses · Procured locally	Above all, a sentimental relationship · Numbers ↘ · 'Love at first sight' horses · Local procurement and imports of horses with particular colours or characteristics
Employment	· ↗↗↗ Employees · ↗↗↗ Self-employed workers Hospitality, instruction, activities, management	· ↘↘ Employees · ↗ Independent instructors Luxury industry training	· ↗ Employees · ↗↗ Volunteers Bringing together multiple skills (psycho-social skills and knowledge of horses, public service and riding, etc.)	· ↘ Jobs in riding lessons · ↘↘ In racehorse training · ↗ Consultants for breeding /natural horsemanship/care and treatment · ↗ Veterinary services
Distribution	· Tourist areas, holiday destinations for urban dwellers · Demographically dynamic regions · Regions breeding horses · Limits: land pressure and transport times in urban and peri-urban areas	· Tourist areas visited by the elite · Areas specialized in breeding high-level horses · Breeding and training migrate to areas with less land pressure	· Varying degrees of development depending on local projects, specificities, and culture	· All rural and peri-urban areas where pasture is available · Sanctuaries for breeds in danger of extinction, or for retired horses at the end of their lives

Key lessons learned from the foresight study

Issues common to all scenarios

Of all the issues common to the scenarios, the most notable is the relationship between humans and horses. It is the key factor influencing the changing role of horses in society.

Questions as the economic efficiency of the industry also figured prominently. In all the scenarios, there are serious concerns regarding the cost of breeding horses and of equestrian services. These costs may be associated with offering more diverse products, improving quality, attracting new customers, being more competitive internationally, remaining accessible as a non-luxury product in difficult economic times, confronting rising food and energy costs, promoting certain – expensive – horse-based activities, caring for aging horses, managing horse health and welfare, dealing with the disposal of dead animals, and dealing with increases to the VAT (value added tax). As it stands, developments in the industry are likely to be highly susceptible to changes in the economic environment and to the availability of more competitive supply from other countries.

All scenarios must deal with the increase of environmental concerns and pressure on resources. The industry does have certain environmental advantages when compared to other herbivores; it produces less greenhouse gas emissions, maintains grassland areas, and contributes to landscape quality. However, there is a real danger that horse breed diversity is being eroded. The current threat to draught horse breeds may also spread to other types of horse since commercial interests do not favour the safeguarding of purebred populations.

Animal health, welfare, and end of life are also of reoccurring concern. The general trend across the scenarios is towards increased medical attention and health monitoring, and better welfare conditions. As regards the risk of new illnesses potentially appearing and spreading throughout France, there was no differentiation among the scenarios, irrespective of the size of the horse population. Horse aging and end-of-life management are major issues that will require specific humane approaches and alternative solutions to slaughterhouses and knackeries (Perrier-Cornet *et al.*, 2013).

Research and development prospects

Above all: improve knowledge transfer

Since the 1970s, public research bodies have amassed considerable knowledge, practices, and tools, however the transfer of this knowledge to users is very much lacking. Given that most horse breeders are individual hobbyists spread across France, it is difficult for information and guidance to reach its intended audience. As a priority, other methods of reading breeders must be devised in order to strengthen the link between research and its beneficiaries. By doing so, new development opportunities can be explored.

Better understand animal behaviour and the human-animal relationship

Particular attention should be paid to understanding animal behaviour and to the human-animal relationship. To this end, temperament analysis would allow calm horses to be identified and used with low-level riders (clubs, trail rides). It could keep skittish or aggressive horse out of contact with people who have little or no knowledge of horses and prevent household accidents with many new horse owners. Temperament analysis could also be used to avoid breeding horses that would find no vocation because of unsuitable temperaments. A better understanding of the relationship between humans and horses could also improve horse training techniques and lead to better technical and economic efficiencies.

Better understand industry stakeholders, users and activities

The study drew attention to the strong need for better technical, economic, and social understanding of equestrian activities in practice, of horse industry stakeholders, and of its end users. Upstream sociology and marketing research is necessary to understand the diverse ideas, uses, expectations, and relationships people have with horses. Broader research in economics, management, and sociology, carried out alongside highly applied research and development projects, is needed to inform debate as to the organization of work in equestrian centres and the production cost of breeding and rearing horses. Research and development programmes must find ways to innovate and improve business competitiveness and adaptability to demand-side imperatives.

Breeding horses adapted to their use

There is clear potential for modern genetic tools to improve the sport performance of animals in competition. However, this potential must also be developed, with a different set of evaluation criteria, to benefit other horse activities as well. In each scenario, the selection of the most suitable, and often highly specialized, genetic profiles facilitates the modernisation and intensification, to varying degrees, of breeding activity. The selection process thus requires a clearer definition of its objectives, increased innovation in terms of tools and techniques, a more cohesive organization of industry stakeholders, and a more widespread acceptance of these technologies.

The industry can become more efficient by encouraging a more focused approach to improve animal genetic value and by maximising a horse's potential by identifying its abilities from a very early age. The importance of wide-scale, early phenotyping must also be underscored, requiring criteria to be defined, tools developed, and objectives validated. Correspondingly, the ability to share trustworthy and impartial information on the quality of breed animals is paramount to maintaining genetic progress.

Improve breeding practices and the environmental impact of horses

As a part of efforts to develop sustainably, a further research priority for horse breeding and rearing should focus on cost-reduction and improving animal welfare. The research should explore, *inter alia*, the ability of horses to maximise the use of less costly feed, of grazed grass, and of dried fodder to meet their nutritional needs over the course of multiple breeding and use cycles. In parallel, programmes should also be developed to evaluate the environmental impact of horses according to breeding practices, their various uses, and their distribution.More broadly, changes to the European Union's Common Agricultural Policy in favour of more environmentally sensitive agriculture practices and support for rural development may spur research into the ecosystem services provided by the horse industry, particularly with regard to the role of horses in preserving grassland areas.

Better manage horse health

Lastly, the scenarios call for increased research into animal health issues. While health, welfare, and performance are, in all scenarios, highly interconnected, the respective importance of each differs. Certain scenarios may place high priority on public health, on user safety (focusing, for example, on developing safety equipment like helmets, safety vests, and guardrails, and on developing teaching methods that ensure rider safety), or on improving performance (through, for example, athlete health and performance monitoring, and health safeguarding). Other scenarios emphasize horse aging issues, the possibility for horses to 'retire', and end-of-life management. In all scenarios, epidemiological surveillance and disease prevention are essential.

Conclusion

Although different parts of the horse industry operate with their own rationales and may even act independently of other industry sectors, and while the scenarios describe highly divergent possible futures, a number of common issues were brought to the fore. They include the place of horses in society, the need to control costs across various levels of the industry, the environmental impact of horses, concerns for animal welfare, and the threat to horse biodiversity.

Generally speaking, analysis of the study's scenarios substantiates current public research on the industry. However, the analysis also highlights the need to restructure the existing hierarchy of research in favour of work in economics, sociology, ecology, and animal behaviour. Scenario analysis also draws attention to the need to re-examine knowledge and innovation transfer processes so that they may be more widely accepted.

Research outcomes provide some flexibility for the industry to respond to possible developments in the future. The work was presented as a part of a dedicated national symposium held on 2 October 2012 and was taken up in numerous articles in veterinary and horse publications. The study's finding were disseminated and discussed in a number of French regions to promote dialogue with industry stakeholders who tend to be very disparate and represented by a multitude of local associations. The next step is to share findings with European horse industry stakeholders to open the way for exchange, not only of research, but also of expertise, which, in the light of EU policy, should be encouraged.

Acknowledgements

The authors wish to thank all of the participants in this study, especially J.-L. Andréani, E. Bour-Poitrinal, J.-L. Bourdy-Dubois, F. Chauvel, I. Ferté, J.-Y. Gauchot, F. Haussherr, E. Heurgon, P. Julienne, L. Lansade, P. Lekeux, G. Mahon, W. Martin-Rosset and B. Morhain – the members of the panel – for their generous and kind involvement.

References

Börjeson, L., M. Höjer, K.-H. Dreborg, T. Ekvall and G. Finnveden, 2006. Scenario types and techniques: towards a user's guide. Futures 38: 723-739.

Boyer, C., 2012. Les infrastructures du tourisme équestre. In: Equi-meeting tourisme, École Nationale de l'Équitation. Mai 9-11, 2012. Saumur, France, pp. 165-166.

Couzy, C., 2007. Peut on encore parler d'éleveur, de cheval ou d'équitation de sport ou de loisir?, 33ème Journée de la recherche équine, Paris, France.

De Jouvenel, H., 2000. A brief methodological guide to scenario building. Technological Forecasting and Social Change 65: 37-48.

Digard, J.-P., 2004. Une histoire du cheval. Art, techniques, société, Arles, France.

Digard, J.-P., L. Ould Ferhat, C. Tourre-Malen, A. Caporal and N. Vialles, 2004. Cultures équestres en crise: Professionnels et usagers du cheval face au changement. Report prepared for the Haras Nationaux, Paris, France, 38 pp.

FranceAgriMer, 2010. La consommation française de viandes. Evolutions depuis 40 ans et dernières tendances, Les synthèses de FranceAgriMer. FranceAgrimer, Montreuil-sous-bois, France, 8 pp.

Godet, M., 2000. The art of scenarios and strategic planning: tools and pitfalls. Technological Forecasting and Social Change 65: 3-22.

Heydemann, P., S. Boyer, C. Couzy, X. Dornier, L. Madeline, B. Morhain and N. Ragot, 2011. Panorama économique de la filière équine, Le Pin-au-haras, France, 241 pp.

Jez, C., B. Coudurier, M. Cressent and F. Méa, 2013a. Factors driving change in the french horse industry to 2030. Advances in Animal Biosciences 4: 66-105.

Jez, C., B. Coudurier, M. Cressent and F. Méa, 2013b. Scenarios. Advances in Animal Biosciences 4: 106-115.

Ministère des sports, Ministère de l'agriculture and Ministère du budget, 2003. Une nouvelle politique pour le cheval. Document de presse, Le Pin-au-Haras, France, 28 pp.

Paillard, S., S. Treyer and B. Dorin, 2010. Agrimonde. Scénarios et défis pour nourrir le monde. Matière à débattre et décider éditions Quae, Versailles, France, 296 pp.

Perrier-Cornet, P., E. Rossier and C. Jez, 2013. Key lessons learnt. Advances in Animal Biosciences 4: 116-123.

REFErences, R., 2012. Filière équine, chiffres clés 2012. Institut français du cheval et de l'équitation.

Réseau REFErences, 2011. Annuaire écus 2011, tableau économique, statistique et graphique du cheval en france, données 2010/2011. Institut Français du cheval et de l'équitation, 63 pp.

Roche, D., 2008. La culture équestre de l'occident xvi-xix siècle. L'ombre du cheval. Tome 1, le cheval moteur, essai sur l'utilité équestre, 479 pp.

Rossier, E., J. Coleou and H. Blanc, 1984. Les effectifs de chevaux en france et dans le monde. In: Le cheval: reproduction, sélection, alimentation, exploitation, Paris, France, pp. 11-24.

Tourre-Malen, C., 2009. Évolution des activités équestres et changement social en france à partir des années 1960. Mouvement Social: 41-59.

Vial, C., M. Aubert and P. Perrier-Cornet, 2011. Les choix organisationnels des propriétaires de chevaux de loisir dans les espaces ruraux. Economie Rurale 321: 42-57.

Viard, J., 2006. Eloge de la mobilité. Essai sur le capital temps libre et la valeur travail, Paris, France.

Wagner, M.-A., 2005. Le cheval dans les croyances germaniques: paganisme, christianisme et traditions. Cahiers de recherches médiévales et humanistes 73: 974.

Part 7.

Horses and sustainable development

Horses have often been blamed for destroying marginal grassland environments by overgrazing and trampling and disturbing the soil. In the next part we feature two reports which are the result of research which demonstrates that, in particular circumstances, horse grazing can improve the biodiversity and productivity of marginal agricultural land. A case is presented to support the use of extensive grazing by horses (sometimes in combination with sheep) to manage the growing amount of European farmland which is being let go to fallow. Here there is an opportunity to use horses to graze these otherwise abandoned lands in a way which keeps the land healthy, and which provides a low cost yet nutritious production base for horses.

15. Sustainable development and equids in rural areas: an open challenge for the territory cohesion

N. Miraglia

Department of Agriculture, Environment, Feedstuffs, Molise University, Via de Sanctis, 86100 Campobasso, Italy; miraglia@unimol.it

Abstract

The European countryside has a great deal to offer; in fact it gives essential raw materials, it represents a place of beauty, rest and recreation and a battleground for the fight against climate change. Many people are attracted by the idea of living and working there. In the past the total number of resident and raised animals decreased with related problems linked to forest and woodlands fires, to soil erosion, to desertification, etc. But over the last years considerable worldwide changes in rural environment have occurred. Europe's rural areas face significant challenges and, between them, the opportunities and capacity to create sustainable development. Central to the idea behind sustainable development is that the long-term preservation of the environment and habitat, as well as its biodiversity and natural resources, will only be possible if combined simultaneously with economic, social and political development particularly geared to the benefit of the poorest members of society. Equids are now playing a rising role in animal biodiversity preservation and in the micro economy of rural areas. Horse husbandry contributes to the diversification of agricultural activities and to the utilization and preservation of extensively cultivated and natural areas, as well as the development of agro tourism on horse farms. In EU's rural areas the diversity of the 'equine culture' and equine-related activities are focused to the leisure and tourism activities, to the preservation of rural socio-cultural life and to the most relevant socio-economic issues. They all represent an open challenge for the territory cohesion. This paper will emphasize the role of equids in the rural development in Europe, identifying the farming systems in a general context of environment and landscape safeguard and of biodiversity preservation.

Keywords: equids, sustainable development, rural areas, territory cohesion, equestrian tourism

The context of rural development in Europe

Based on population density, all rural areas represented 91% of the territory and 55% of the population of the EU-27 in 2007. The corresponding shares for *predominantly rural* areas alone were 69% of the territory and 24% of the population making them particularly important in terms of land use (EU Commission, 2009).

Rural areas generated 48% of the total GVA (gross value added) and provided 56% of the overall employment (EU Commission, 2010, 2013). The Rural development policy regulation for 2007 to 2013 concerns three thematic axes focused to improve the competitiveness of the agricultural and forestry sector, the environment and the countryside, the quality of life in rural areas, encouraging diversification of the rural economy.

The Agricultural Policy is focused on global management and on the promotion of sustainable agriculture to respond to social and economic needs. More recently, the Overview of CAP Reform 2014-2020 (EU Commission Policy 2013-2020) continues along this reform path, moving from product to producer support and now to a more land-based approach (Basic regulation of the new CAP, 2013). This is in response to the challenges facing the sector, many of which are driven by factors that are external to agriculture. These have been identified as economic (including food security and globalization, a declining rate of productivity growth, price volatility, pressures on production costs due to high input prices and the deteriorating position of farmers in the food supply chain), environmental (relating to resource efficiency, soil and water quality and threats to habitats and biodiversity) and social (where rural areas are faced with demographic, economic and social developments including depopulation and relocation of businesses).

The attainment of sustainable development implies the balance between these three objectives and their simultaneous achievement (Nijkamp, 1990; Dourojeanni, 1993). Agriculture makes a valuable contribution to the socio-economic development of rural areas and a full realization of their growth potential (Boyazoglu, 1998). In this context the rural development is a vitally important policy area inside the EU priorities (EU CAP IInd Pillar). The diversification of the economy of rural areas is linked to sectors other than agriculture. As a result, in 2007, 86% of employment and 95% of value added in predominantly rural areas of the EU-27 came from the non-agricultural sectors. Among these, tourism is one of the key opportunities in terms of potential growth for rural areas (EU rural development, 2006, 2009, 2010, 2013).

Agriculture and forestry are the main land users and play a key role in the management of natural resources in rural areas and landscape. Europe's citizens are deeply attached to the diversity of landscape created by the wide variety of agricultural structures and farming types in the EU. Safeguarding the rural environment means investing in the future, creating new employment possibilities and encouraging rural diversification. People must be offered opportunities to create wealth as well as long-term rewarding job prospects. These elements represent a strong advantage in the scope of the new deal for territories management (Miraglia and Martin-Rosset, 2013; Miraglia et al., 2006, 2010).

The management of natural resources in the EU's rural areas represent a platform for economic diversification in rural communities because they offer real opportunities in terms of potential for growth in new sectors, provision of rural amenities and tourism, attractiveness as a place to live and work, reservoir of natural resources and highly valued landscapes (Miraglia, 2014). The valorisation of all agro touristic resources represents significant economic advantages coming from a touristic offer competitive with other activities associated to sustainable agriculture.

Horse and territory: building a new rural development

Rural development is a key tool for the restructuring of the agriculture sector encouraging diversification and innovation. Europe's rural areas are diverse and include many leading regions. In areas designated as 'less-favoured', agricultural production and related activities are more difficult because of natural handicaps, e.g. difficult climatic conditions, steep slopes in mountain areas, or low soil productivity (Miraglia et al., 2010).

It is estimated that high nature value farmland covers more than 20% of agricultural area in most Member States (even more than 30% in some of them). More generally, 16% of the EU-27 utilized agricultural area is located in mountainous areas, where agriculture contributes actively to maintaining biodiversity. Appropriate methods of production, such as extensive farming, may also support biodiversity (EU Commission, 2009).

In term of preservation of environment, landscape, biodiversity of nature and output from grassland, it has been estimated that, in Europe, grass and preserved forages represent 50-80% of horse's feeds in the year; the grazing activity is between 6 to 10 months, depending on the European country, and forages can supply 40-70% of annual feeds requirements of the horse. Actually 3.2% of the agricultural surface is destined to horse feeding during the year and it has been estimated that across Europe, approximately 6.8 million hectares of land are used to breed horses (EPMA, 2009).

The rural development of marginal lands with the breeding of equids represents many potential economic advantages coming from different kinds of production, from the improving of sustainable agriculture models, and from the increase of tourism demand for the discovery of mountain and hilly areas. They represent and add value to land, maintaining the rural landscape and the biodiversity of flora and entomo-fauna, and maintaining population and socioeconomic activities in rural areas. Resources are structural elements in the challenges posed by territory building and traditional products. In 2006, Casabianca and Matassino (Casabianca and Matassino, 2006), using two interesting 'Mandala' exemplification models, showed the diversified components of 'territory' with their interrelationship and interdependence ratios together with the factors influencing the 'typified traditional' products. Many activities are expression of all these aspects.

In terms of horse use, the 'Main activities' depend on the more traditional use of horses: races, sports, leisure and hobby farming (IOC, 2009); the 'New activities' include Socio-cultural events, Agritourism, Medical therapy, Social rehabilitation and 'Green care activities'; the Niche activities concern the products, meat and milk (Miraglia and Martin-Rosset, 2013; Miraglia et al., 2010). Local traditions influence the agricultural choices and also the equids production that are often directed toward the exploitation of marginal lands.

Recently there has been a growing interest directed toward some autochthonous horse populations destined to saddle, meat and milk production; toward mules that are more and more requested for work; and toward asses that are used for milk production mainly directed to infant nutrition.

These local equine breeds represent special niches strictly linked to socio-cultural and economic activities. Many of these breeds (e.g. Lipizzan horses, Old Kladrub horse, native horse and pony breeds in Britain, Connemara Pony populations, Croatian autochthonous horse breeds, horse breeds in Hungary, Akhal-Teke horse breed in Turkmenistan, rare horse breeds in Northern Europe, endangered Norwegian horse breeds, Italian Bardigiano horses, Italian Heavy Draught Horse breed, Pentro horses, Retuertas horse, Zemaitukai Horse and many others) have an important role in the safeguarding of local biodiversity and can be considered a real resource because it expresses variety of genetic information and is thus able to flourish in a variety of ecosystems. In Europe we find at least 2 wild populations that we can consider really 'feral horses': the Wild Horses in Romania that found their habitat in the Donau delta which represents a real paradise for flora and fauna and the Wild Horses in Germany, the Dülmen Pony, who are totally able to live on their own in the wild nature. In times of strong rain or an intense solar radiation the animals do have a refuge underneath the treetops. Moreover, it exists other situations of European wild horse populations bred on free-ranging feral or semi-feral conditions such as Sorraia in Portugal, Konik in Poland, Pentro in Italy; in most part of these cases there are biodiversity programs in relation to the landscape safeguard and sustainable development (Miraglia et al., 2008). Many of these breeds occupy special niches and contribute to the biodiversity due to their own genetic characteristics coming from adaptation mechanisms developed in centuries of evolution in specific local environments (Edouard et al., 2008; Fleurance et al., 2001). The EEC Global Strategy for the Management of Farm Animal Genetic Resources of FAO (1998 and 2010) organized specific programs for the management of genetic resources at local, regional, national and international level. As a consequence some policies concerning the safeguard of endangered breeds and of autochthonous populations were developed. They are focused also on the recovery of the relationships of 'human-animal-territory' which could be considered as a 'system integrator' of all the processes concerning the rural eco-sustainable development (Van der Zijpp et al., 1993; Yachi and Loreau, 1999) including potential economic advantages from the different kinds of production.

The autochthonous populations and their preservation also represent an important issue in the context of milk and meat products. Mares' and ass's milk has in fact stimulated much interest in the last years (Salimei, 2010). Equine milk is regarded as having nutritional and therapeutic properties which are beneficial to the elderly, convalescents and infants. Similar to human milk, it is low in protein, has low casein to whey protein ratio, high lactose, a high concentration of polyunsaturated fatty acids and low cholesterol. Equine milk and products there from (e.g. koumiss) have always been used widely in Russia (for the treatment of tuberculosis) and Mongolia. Estimates suggest that more than 30 million people worldwide drink equine milk regularly, and that figure has increased significantly in recent years. Currently, there is considerable interest in the use of equine milk in human nutrition in western Europe where supply now falls short of demand. A decade ago, equine milk was produced only in isolated small holdings in Eastern Europe, Belarus, the Ukraine and Mongolia but now there are large-scale operations in France, Belgium, Germany, Italy, Austria and the Netherlands. Asinine milk for human consumption is mainly in areas where donkeys are traditionally bred, Asia, Africa and Eastern Europe but consumption in Western European countries is increasing due, as with equine milk, to the similarity of its composition to human milk. Both milks are considered good substitutes for bovine milk for children with severe IgE-mediated cow's milk allergy. The high lactose of these equid milks ensures good palatability and improves absorption of calcium, which is essential for bone mineralization in children. Equid's milk for human consumption could contribute to the animal biodiversity preservation but could also positively influence the micro-economy of marginal and hilly areas, where the total number of residents and raised animals are decreasing, with related social problems of forest fire, soil erosion and desertification (Miraglia *et al.*, 2003).

At the same time equids have evolved to address new social roles: the use of equines in human rehabilitation and hippo therapy, these can have a positive role in the treatment of different kind of psychotic pathologies. They include recreational, educational, or therapeutic goals achieved through relationship with horses. They support the social growth and welfare of 'at-risk' children, youth and adults (Kjäldman, 2010).

Rural development and equids: a challenging issue to encourage diversification and innovation

Horse breeding in rural areas represents an essential element of landscape and is located often where it is difficult to develop traditional agricultural activities. The equine industry is facing a rising socio-economic demand whilst the production and utilisation of equines is much diversified. This evolution is supported by many Western European countries because horses are part of territory management (Miraglia and Martin-Rosset, 2013).

Recent growth sectors include: tourism, crafts and the provision of rural amenities. These elements constitute real opportunities for on-farm diversification outside agriculture and the potential development of micro businesses in the broader rural economy. The preservation of socio-cultural traditions and the diversification of the activities linked to horses play an important role in the valorisation of territories (Miraglia, 2014; Miraglia *et al.*, 2006). The link is often represented by the riders attracted to equestrian promenades – which occur in 70 to 80% in industrialized countries.

Socio-economic aspects and equestrian tourism

Equestrian tourism is now beginning to be professionally organised mainly in those Western European districts where the 'horse culture' is more developed: many traditional horse events highlight the rural scene and the equestrian activities are linked to typical horse breeds. Many economic advantages are connected to horse activities: and their presence helps safeguard nature and cultural heritage. In this way, horses could become the mediator between economic and socio-cultural aspects.

The development of agro tourism in Europe is linked to the valorisation of the environment and represents a considerable source of non-material values and image. The most important objectives of the 'Common Strategic Framework for rural development' of CAP 2014-2020 is the valorisation of the local balance and of social dynamics *via* the rural tourism, to start new economic activities and to create new opportunities in rural communities. In this context, the development of tourism is integrated in a global market – 'think global – act local'.

The success in this field is to add new ideas to the scope of Rural Tourism and to protect the rural communities and their typical traditions. It represents an open challenge for territorial cohesion and for the reduction of the poverty in those regions.

A possible European program which could be used to promote and manage rural tourism could be created by involving the project in the development of rural tourism and of Green Ways (www. greenways.by). Joining agro touristic and equestrian activities may represent a real opportunity, and an advantage for a farm in the local market. Such activities could be managed by local people in order to encourage sustainable development and healthy lifestyle, and in integrated manner which enhances both the environment and quality of life of the surrounding area.

This is why equestrian tourism is now the main 'knowledge-activity' of the recreational equestrian sport. It is a field which does not imply a sustained activity, but on the contrary, a sporadic activity that allows riders to take delight in riding the trails in nature when time and financial resource permit.

Equestrian tourism farms offer the opportunity to enjoy horses and proper equipment, offering various recreational activities in the surroundings on horseback. The equids commodity chain generates employment and job at different levels and at different typologies. These different tasks are carried out by people with different expertise (equine veterinarians, horse nutritionists, farriers, riders, grooms, horse breaker, head lead, etc.) and specialised industries (equine veterinary products, horse clinic, specialised feeds and supplements, artificial insemination, organisation, societies, etc.) that represent direct profits. Other profits that depend directly on equid's breeding come from specialised horse equipment, rider equipment, services companies, specialised press, specialised corporations, equestrian show and promenades, riding schools and betting. In this way, a wide productive and trading mechanism is developed (Miraglia *et al.,* 2006). In the case of equestrian tourism the profits are linked also to the typology, because it is often a seasonal and complementary activity. Moreover, in some cases the professionals are breeders and not professional of tourism and don't organize their activity to make profit enough.

Preservation of traditional socio-cultural life

Equids represent a strong positive imaginary for people because of cultural heritage and strong historical links. Involving equine breeds and geographical areas can create a local trademark image valorised by tourism. The most interesting cases in Europe can be found in the following countries. In France: the Camargue regional natural Park, the Marquenterre Park, the Perche regional natural Park, the Loire Anjou Valley regional natural Park; in UK: the New Forest National Park; the Yorkshire Dales National Park, the Scottish Islands; in Norway: the Dombas National Park; in Poland: the Bieszczadzki National Park; in Czech Republic: the Kladubry National Park; in Hungary: the Aggtelek National Park and the Hotobagi National Park; in Spain: the Menorca Islands; in Italy: the Pentro Horse Park, the Maremma district, the Nebrodi Park and the Giara district; in Portugal: the Peneda-Geres National Park (Miraglia *et al.,* 2006).

Concerns exist that the relationship between a citizen and cultural rural life is becoming more and more diffused. However, many equestrian events are attended by thousands of people and represent a strong positive image because of the cultural heritage and strong historical links (Miraglia *et al.,* 2006: Miraglia *et al.,* 2010). In this way agritourism offers the possibility to welcome tourists together with a strong stimulation to maintain services and events.

The equestrian tourism is beginning to become professionally organised across Europe. The preservation of traditional socio-cultural life tends to support and mobilize local communities – encouraging local enterprise, creating jobs and additional revenue streams, restoring and protecting traditional vocations. This means the use of local resources, accommodation and food, tourist services and local products. These aspects help local communities to discover and strengthen their cultural and social identity, improving conditions and quality of life. At the same time they provide information and opportunities for tourists to help them to better understand the region, its challenges and local initiatives, activities and organizations.

Conclusion

Equids and the socio-economic development of the sector offer both direct and indirect profits coming from a full-fledged economic sector based on employment and job, quality of products, and the development of breeding strategies in the context of sustainable agriculture. Yet it is seldom that equine tourism businesses can exist outside of a broad suite of farm income strategies. By developing the equine tourism sector, more economic activity can be generated by what may be otherwise marginal farmsteds.

Actually the equid sector matches the most important expectations coming from modern society, because they are involved on one side in the preservation of the environment, landscape and output from grassland and the preservation of livestock biodiversity and, on the other side, in sustaining the relationship between citizens and cultural rural life and in the preservation of traditional socio-cultural life. In these situations equids contribute to the preservation of the socio-cultural traditions, to the development of equestrian tourism and, consequently, to the valorisation of the horse in rural areas that stimulate also socio-economic relationship between urban and rural areas.

In synthesis, the valorisation of all the agritourist resources represent significant economic advantages coming from a touristic product competitive with other activities associated with sustainable agriculture. The potential economic advantages come from different kind of equid productions, from the improvement of sustainable agriculture models and from the increase of tourism demand for the discovery of mountain and hilly areas. These aspects are considered of primary importance in the new EU CAP policy (CAP 1st pillar and 2nd pillar). In this context the main achievements and the new prospects are represented by technical and socio economic evaluation of the chain components: sectors, structures, territories activities and jobs, all involved in creating new opportunities to manage and capitalize on territorial management (Miraglia, 2014; Miraglia and Martin-Rosset, 2013; Miraglia *et al.*, 2006, 2010).

References

Boyazoglu, J., 1998. Livestock farming as a factor of environmental, social and economic stability with special reference to research. Proceedings of 36th Congress of the South African Society of Animal Science 'Animal Production in harmony with the environment'. University of Stellenbosch and Elsemburg Agricultural College, South Africa.

Casabianca, F. and D. Matassino, 2006. Local resources and typical animal products. In: Rubino R., L. Sepe, A. Dimitriadou and A. Gibon (eds.) Product quality based on local resources leading to improve sustainability. EAAP Scientific Series no. 118. Wageningen Academic Publishers, Wageningen, the Netherlands, pp. 9-26.

Dourojeanni, A., 1993. Procedimientos de gestión para el desarrollo sustentable: aplicados a microrregiones y cuencas. Instituto Latinoamericano y del Caribe de Planificación Económica y Social de las Naciones Unidas (ILPES). Santiago, Chile.

Edouard, N., G. Fleurance, W. Martin-Rosset, P. Duncan, J.P. Dulphy, H. Dubroeucq, S. Grange, R. Baumont, F.J. Perez-Barberia and I.J. Gordon, 2008. Voluntary intake and digestibility in horses: effect of forage quality with emphasis for individual variability. Animal 2(10): 1526-1533.

European Commission, 2009. Commission communication: 'towards a better targeting of the aid to farmers in areas with natural handicaps'. Available at: http://ec.europa.eu/agriculture/rurdev/lfa/comm/index_en.htm.

European Commission for Agriculture and rural development, 2010. Situation and prospects for EU agriculture and rural areas. Available at: http://tinyurl.com/oqya2y9.

European Commission, 2013. Overview of CAP reform 2014-2020. Agricultural Policy perspectives briefing 5. Available at: http://tinyurl.com/nmwkrdx.

EU Rural Development in the European Union, 2006. Community strategic guidelines for rural development. Available at: http://tinyurl.com/o4ee593.

EU Rural Development in the European Union, 2009. The overview of rural development policy 2007-2013. Available at: http://ec.europa.eu/agriculture/rurdev/leg/index_en.htm.

EU Rural Development in the European Union, 2010. Rural development in the European Union – statistical and economic information. Available at: http://ec.europa.eu/agriculture/agrista/rurdev2010/ruraldev.htm.

EU Rural Development in the European Union, 2013. Rural development in the European Union – statistical and economic information. Available at: http://tinyurl.com/npbohqb.

European Pari Mutuel Association (EPMA), 2009. The economic and social contribution of horseracing in Europe. Available at: www.parimutuel-europe.org.

FAO, 1998. Primary guidelines for development of national animal genetic resources management plans. Available at: http://tinyurl.com/onfa9f9.

FAO, 2010. Funding strategy for the implementation of the global plan of action for animal genetic resources. Available at: http://tinyurl.com/peq45uy.

Fleurance, G., P. Duncan and B. Mallevaud, 2001. Daily intake and the selection of feeding sites by horses in heterogeneous wet grasslands. Animal Research 50(2): 149-156.

International Olympic Committee, 2009. Guide on sport, environment and sustainable development. Olympic Studies Centre, Lausanne, Switzerland.

Kjäldman, R., 2010. The impact of socio-pedagogic equine-activities. Intervention on special education pupils with neurological disorders. In: EAAP book of abstracts no. 16. Wageningen Academic Publishers, Wageningen, the Netherlands, p. 343.

Miraglia, N., 2014. The horse as a key player of rural development in Europe. In: EAAP book of abstracts no. 20. Wageningen Academic Publishers, Wageningen, the Netherlands, p. 269.

Miraglia, N. and W. Martin-Rosset, 2013. Utilisation du territoire par le cheval au niveau européen: acquis et perspectives dans le cadre des activités de la Commission Equine de la Fédération Européenne de Zootechnie. 39th Journée de la Recherche Equine. February 28. Paris, France.

Miraglia, N., M. Polidori and E. Salimei, 2003. A review on feeding strategies, feeds and management of equines in central-southern Italy. In: Pearson, R.A., P. Lhoste, M. Saastamoinen and W. Martin-Rosset (eds.) Working animals in agriculture and transport. EAAP Technical Series no. 6. Wageningen Academic Publishers, Wageningen, the Netherlands, pp. 103-112.

Miraglia, N., D. Burger, M. Kapron, J. Flanagan, B. Langlois and W. Martin-Rosset, 2006. Local animal resources and products in sustainable development: role and potential of equids. In: Rubino R., L. Sepe, A. Dimitriadou and A. Gibon (eds.) Product quality based on local resources leading to improve sustainability. EAAP Scientific Series no. 118. Wageningen Academic Publishers, Wageningen, the Netherlands, pp. 217-233.

Miraglia, N., M. Costantini, M. Polidori, G. Meineri and P.G. Peiretti, 2008. Exploitation of a natural pasture by wild horses: comparison between nutritive characteristics of the land and the nutrient requirements of the herds over a 2-year period. Animal 2(3): 410-418.

Miraglia, N, M. Saastamoinen and W. Martin-Rosset, 2010. The role and potential of equines in a sustainable rural development in Europe – social aspects. In: EAAP book of abstracts no. 16. Wageningen Academic Publishers, Wageningen, the Netherlands, p. 339.

Nijkamp, P., 1990. Regional sustainable development and natural resource use. In: World Bank Annual Conference on Development Economics. Washington, DC, USA.

Salimei, E., 2010. Advances on Equus asinus as a dairy species. In: EAAP book of abstracts no. 16. Wageningen Academic Publishers, Wageningen, the Netherlands, p. 42.

Van der Zijpp, A., J. Boyazoglu, J. Renaud and C. Hoste (eds.), 1993. Research strategy for animal production in Europe in the 21st century. EAAP Publication no. 64. Wageningen Academic Publishers, Wageningen, the Netherlands, 163 pp.

Yachi, S. and M. Loreau, 1999. Biodiversity and ecosystem productivity in a fluctuating environment: the insurance hypothesis. Proceedings national academy of sciences 96(4): 1463.

16. Roles of horses on farm sustainability in different French grassland regions

G. Bigot[1]*, S. Mugnier[2,3], G. Brétière[1], C. Gaillard[2,3] and S. Ingrand[3]

[1]Irstea, 9 avenue Blaise-Pascal, CS 20085, 63178 Aubière, France; genevieve.bigot@irstea.fr
[2]Agrosup Dijon, Bureau S 139, Combe Berthaux, BP 87999, 26 boulevard Docteur Petitjean, 21079 Dijon Cedex, France
[3]INRA Theix, 63122 Saint Genes Champanelle, France

Abstract

In France, horse rearing is often associated with other farming activities such as beef or dairy productions that prevail in grassland areas. To analyse the role of horses on the sustainability of farming systems, we surveyed one hundred farms in four major areas of horse breeding, chosen in regard to their production system representative of regional agricultural systems. The regions differed in their agro-climatic context, associated with a type of horse production: saddle horses in oceanic and continental lowlands, and draught horses in two upland regions in Central and Eastern France. Farmers were questioned about the role of their equine production in the economic, environmental and social functioning of their farming system. Results showed that horse numbers ranged from less than 10 to 100% of the total livestock (expressed in livestock units). In upland areas, draught horses were appreciated for their use of grasslands along with or after cattle. These hardy horses needed little labour, but the low income from this production limited its development. In lowlands, farmers raised saddle horses either alone or with dairy or beef cattle. In farms specialized in horses, breeders developed allied services such as taking horses at livery. In mixed herds, farmers spent comparatively more time on saddle horses than on cattle, whether beef or dairy. The impact of saddle horses on gross production depended on numbers and age for sale. In conclusion, horse rearing presented a low profitability, especially for draught horses, or an unpredictable profitability, especially for saddle horses. However, whatever the types of animal production, horse grazing improved the maintenance of grassland areas, and farmers raised horses, which attracted volunteer labour, because they enjoyed taking care of them.

Keywords: horses, farms, sustainability, grasslands

Introduction

In France, the number of horses has been estimated at nearly 1 million, counting all breeds, breeding and active horses (REFErences, 2011). Nearly half of this livestock is held in farms (Perrot et al., 2013); the rest is kept by individual owners or in other types of organizations. The density of farms with horses depends on region, with a higher concentration in lowland and upland grassland areas (Bigot et al., 2012b). However, in these areas, mainly used by cattle farming, alternative agricultural productions are limited. In particular, sustainability in these zones involves maintaining a rural workforce, based on a viable agricultural economy, while preserving environmental resources. Hence horse farming, supported by the growth of equestrian activities, could provide new opportunities for sustainable development by developing economic and leisure activities in the conservation of a sound environment (Miraglia, 2012).

Accordingly, we set out to assess the effect of horse breeding (alone or associated with cattle) on different sustainability parameters of farms in various grassland regions. This study specifically addressed the place of horse breeding in farming. For this purpose, surveys were conducted in 100 professional structures representative of grassland systems, where production constraints could vary depending on agro-climatic conditions and production targets, including animal husbandry. Here we present the main results on the operating strategies implemented by farmers to ensure sustainability of farm structures.

Materials and methods

Criteria for the selection of grassland regions were based on the size of the horse stock, the type of horse production, and the diversity of the agro-economic context. We thus chose two uplands: Auvergne and Franche-Comté, with much draught horse breeding, and two lowlands: Basse-Normandie and Centre, with much saddle horse farming, but differing in their agro-climatic contexts (Figure 1). Auvergne is the foremost French area for the number of draught broodmares (REFErences, 2013). This upland area supports mainly cattle, third to produce milk for locally made cheeses, and two-thirds for beef production of weanling calves (6 to 10 months) exported to Italy to be fattened (Agreste Auvergne, 2012). Franche-Comté is the birthplace of the horse breed 'Comtois', which became the largest draught breed all over France (REFErences, 2013). Franche-Comté has mainly been devoted to dairy cattle for production of PDO cheeses, based on grazing systems (Agreste Franche-Comté, 2012). Basse-Normandie is the foremost national area for horse numbers, produced particularly for show jumping and for trot and gallop races (REFErences, 2013). This region is also dominated by dairy cattle farming, and secondarily by suckling cattle breeding for grass-fed beef production (Agreste Basse-Normandie, 2012). Centre is an area of mixed farming combining beef cattle and crops, and where a saddle horse breeding for sports has been developed in these agricultural structures (REFErences 2013).

Criteria for selecting farms were: more than three broodmares held, professional status (i.e. more than one working unit), and a farming system representative of each zone.

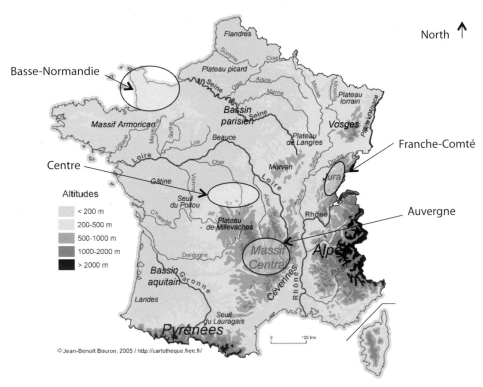

Figure 1. Location of studied areas.

In each farm, a face-to-face interview was carried out with similar questions independently of area. They covered farm structure, land use, number and types of livestock, workforce, economic products, general management, and specifically impact of equine activities in terms of use of agricultural area, contribution to gross product and mobilization of workforce.

To appraise sustainability of each system, we retained one criterion on each pillar: environment, social and economic. For the environment pillar, we retained grazed grassland areas (expressed in hectares) due to the impact of different species grazing on biodiversity (Rook *et al.*, 2004). The workforce (expressed in working units) was the quantitative criterion retained for the social pillar. For the economic pillar, the gross output (expressed in kilo-euros) was retained. These both criteria are often used in economic farm models (Janssen and Van Ittersum, 2007). Thus the impact of horse production on farm sustainability was expressed as a proportion of the total farm performance in each criterion, depending on the horse numbers in the total livestock.

Given the small numbers of farms per system and area (Table 1), qualitative analyses were performed to present main strategies in regard to the farm contexts.

Results

Features of surveyed farms

In each area, 20 to 26 farms representative of their regional systems were surveyed. The results given below concern only farms presenting a grassland area greater than 50% of the total area (Table 1). Surveyed upland farms were between 700 and 1000 meters above sea level, lowland farms lower than 500 m. Farm areas were greater in uplands and in 'horse and dairy cattle' farms from Basse-Normandie than their regional averages, which were close to those of 'horse-specialized' farms. Livestock number varied with farm size and region. In particular, the total stocking rate ranged from 1.0 livestock unit (LU)/hectare (ha) in uplands and in mixed farms in Centre to 1.5 LU/ha in horse and cattle farms in Basse-Normandie. Horse-specialized farms in both lowlands presented a stocking rate 30% lower than farms associating horses and cattle. The total output of surveyed farms in Basse-Normandie was greater than in upland farms with dairy cattle and the total output of farms located in Centre was still lower. The same trends were observed for the average workforce in these different farm groups. The proportion of horse livestock per total livestock was under 15% in draught horse farming in uplands, with little variations according to cattle production: from 7-8% in dairy and beef systems to 13% in dairy systems in both regions. This proportion varied in saddle horse farming in lowlands, from 28% with dairy cattle, to 40% or 46% with beef cattle, and 100% in horse-specialized farms.

Horses and farm sustainability

To analyse the impact of equine production on farm sustainability, we retained only types which included at least 5 farms.

Draught horse farming in uplands

Whatever the orientation of cattle production, draught horses grazed all pastures (Figure 2). According to the farm, they generally grazed plots simultaneously with suckling cows and heifers, but after dairy cow grazing (Bigot *et al.*, 2012a; Mugnier *et al.*, 2014). In particular, in dairy systems where grass quality was important to maintain milk production, the horse grazing helped to remove grasses refused by milked cows. This fact could explain the highest proportion of draught horses in dairy systems in comparison with beef systems where grass quality was relatively less important. Draught horses also grazed plots unusable by cattle because they were too small, too far from stables or had

Table 1. Specifications of surveyed farms for (A) draught horse farming in uplands and (B) saddle horse farming in lowlands.

A. Draught horse farming in uplands	Auvergne			Franche-Comté	
	Horses and beef cattle	Horses and dairy cattle	Horses and both dairy and beef cattle	Horses and dairy cattle	Horses and both dairy and beef cattle
Farm number	8	11	6	22	4
Agricultural area average (ha)	129	128	127	120	199
Permanent grassland (% total area)	79	91	85	82	84
Total livestock average (LU)	135	114	147	105	165
Total output average (k€)	107	175	155	175*	220*
Total labour force average (WU)	1.7	2.5	2.7	2.1	2.3
Rate of horse livestock on total livestock (%)	10	13	7	13	8

B. Saddle horse farming in lowlands	Basse-Normandie			Centre	
	Horses and beef cattle	Horses and dairy cattle	Horse specialized	Horses and beef cattle	Horse specialized
Farm number	6	9	5	7	8
Agricultural area average (ha)	76	128	66	75	49
Permanent grassland (% total area)	83	60	90	88	100
Total livestock average (LU)	109	198	68	73	37
Total output average (k€)	233	345	214	66	85
Total labour force average (WU)	2.6	3.6	3.6	1.4	2.2
Rate of horse livestock on total livestock (%)	46	28	98	40	100

* Estimation.

poor forage value (Bigot *et al.*, 2013). Accordingly, draught horses were also present in beef and dairy cattle farming in uplands but in lower proportion.

On the other hand, the equine herd contributed little to output and less than the horse numbers might suggest, because the main production was the sale of foals just after weaning. However, though small, this contribution to the output required a low mobilization of the workforce especially in Auvergne. In this area, draught horses were raised very extensively: outdoors throughout the year, fed with roughage only during snowy periods, rarely supplemented and reproduced by running service, with no variations remained between cattle systems. In Franche-Comté, work devoted to

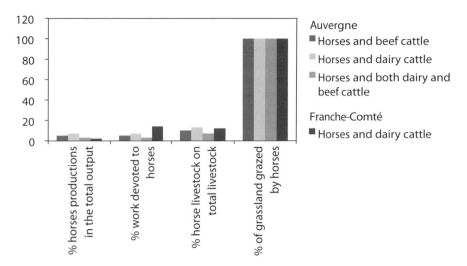

Figure 2. Impact of draught horses production on farm sustainability criteria in regard to cattle production in Auvergne and Franche-Comté.

horses was relatively more important because some farmers housed horses in stable during winter and trained some of young horses for draught activities.

When farmers were asked about their motivation to hold horses, they first answered that they liked the animal, but 85% of them also mentioned its impact on grassland maintenance. In particular, 85% cited the horse's ability to consume grasses refused by cattle, and 40% cited its ability to maintain plots unusable by cattle. In the absence of horses, 44% of surveyed farmers said they would have to use the roller chopper more often, especially in beef systems where the stocking rate was around 1 LU/ha. Only 24% would increase the cattle numbers and 15% would finally abandon maintenance of small plots. Farmers did not want to increase the horse numbers particularly for economic reasons. No breeder presented the economic impact of horses as a motivation for their presence, and 22% of respondents noted the advantage of its low work requirements, particularly in structures with beef and dairy cattle productions.

Saddle horse farming in lowlands

In farms with horses and cattle, impact of horses on farm sustainability criteria was analysed depending on the relative numbers of horses and the cattle system (Figure 3). Firstly, horses and cattle grazed almost all the pasture area with the same combinations as in draught horse farming, irrespective of the proportion of horse livestock in the total livestock. In most farms, mares and young horses up to the age of 3 years were kept outdoors all year. Stallions and horses kept for training to be sold between ages 4 and 7 years were housed in stables and conducted daily in paddocks of exercises that could amount 10-30% of the agricultural area (Bigot *et al.*, 2011, 2012b). These horses did not interfere in the grazing management.

When horses were associated with beef cattle at 40-46%, they contributed as cattle to the total output, and they needed relatively more work time than cattle required in both regions. In dairy cattle systems in Basse-Normandie, the contribution of horses to the farm output was in proportion to their numbers, but saddle horses mobilized the same work as the dairy herd, which was already demanding.

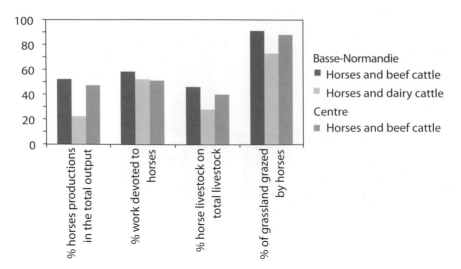

Figure 3. Impact of saddle horse production on farm sustainability criteria in horse and cattle systems in Basse-Normandie and Centre.

This large work demand for saddle horse production could be explained by two main reasons:

1. 75% of the surveyed farms sold young horses aged between 4 and 7 years, i.e. after breaking and training for show jumping which needed individual work for horses instead of more collective work for cattle, even for dairy cattle. In particular, these young horses were housed in individual stalls, and were led to paddock every day. The training also needed trips to competition venues.
2. Most of the surveyed farms proposed services to private owners or other breeders of saddle horses, such as livery boarding, reproduction activities, or training services (Table 2). These activities needed further labour, either hired or family, and also used local volunteers. These activities were more developed in Basse-Normandie, even in dairy cattle farms than in Centre and in horse-specialized farms, irrespective of region. In both regions, all horse-specialized farms provided at least livery boarding.

Discussion

These results provided initial elements on the role of horse breeding in agricultural systems in combination with other equine activities or other animal productions in some various agro-climatic contexts, and socio-economic strategies of farmers.

In our sample, draught horses (expressed in LU) represented a small proportion of the total livestock (less than 15%) whichever the region and in both dairy and beef production. In saddle stud farms, the proportion of horses in the total livestock could vary greatly, in regard to the structures and the bovine specialization. These results were representative of the place of draught and saddle horses in French farms as revealed by different analyses of the national agricultural census (Bigot *et al.*, 2010; Perrot *et al.*, 2013).

Even if interviewed farmers explained the reason of horse breeding firstly because they were attracted to this animal, we found three main types of horse farming with various impacts on economic, social and environmental criteria:

Table 2. Proportion of saddle stud farms proposing services linked to horses according to region and cattle production.

Services (% of the farms number)	Basse-Normandie			Centre	
	Horses and beef cattle	Horses and dairy cattle	Horse specialized	Horses and beef cattle	Horse specialized
Farms number	6	9	5	7	8
Livery boarding	67	33	100	29	100
Reproduction centre	33	11	80	14	50
Training centre	50	22	60	0	0

The first category of farms ensured their economic resources to productions other than horses. This was the case for professional farms breeding draught horses in uplands like Auvergne and Franche-Comté. The horses, present in small numbers compared with cattle, supplemented bovine herds for better utilization of grasslands. No substantial income was expected, but horses indirectly contributed to a greater efficiency of the forage system with a very light mobilization of the workforce. In this first category could also be integrated farms where saddle horses represented less than 30% of the total livestock, such as farms with dairy cattle in Basse-Normandie. In these cases, farmers restricted the equine herd to availability of grassland surfaces and/or work force, and set the profitability of horse production as best they could. In both these cases, horse activities could even be considered as a 'hobby' by farmers.

The second category of farms associated saddle horses with a comparable number of beef cattle. In all these farms, grasslands represented more than 80% of the agricultural area. The presence of the two herbivores maximized the use of grasslands, even those with poor forage value. In these structures, breeders spent relatively more time with their horses than with their cattle. The equine output was a bit superior to the output of beef productions including their specific subsidies, but cattle production ensured a relatively stable income that compensated for income fluctuations due to late sales of horses after training.

The third category of farms was specialized in saddle horse production. In these farms, the use of grasslands was not as efficient as in systems associating horses and cattle. These results could be explained in two ways: (1) the specific patterns of grassland use made by horses: over- and under-grazing of areas (Fleurance *et al.*, 2012); or (2) a more extensive management of grasslands, because available workforces were deliberately more directed to the horse handling than to the maintenance of forage areas. All these structures developed a complementary activity (reproductive services, or boarding horses with or without work) with a view to improving income. These activities answered to local or more distant demands of breeders or private horse owners. In providing services and jobs, these farms had a meaningful social impact at the local level.

Conclusion

These results from farms in different contexts highlighted the main incidences of horse breeding on sustainability of grassland farming. First an environmental impact: in all these farms, breeding horses contributed to the use and maintenance of large pasture areas by their grazing and in broader ways than cattle. Whatever the orientation of livestock productions: saddle or draught horse, dairy or beef cattle, the association of the two species contributed significantly to a best use of grasslands and to a best efficiency of farming systems. Horse breeding also contributed to animal biodiversity:

French horse numbers amount to only 5% of bovine numbers, and all French draught horses are considered endangered animals due to their small numbers in each breed. Secondly, the economic impact seemed related obviously to the horse number but also to the type of horse production (more moderate in draught breeds than in saddle productions) and to the presence of services related to horses. Finally, the social impact was more important than the only breeder satisfaction. Draught horse breeding in mountains simplified the mechanical maintenance of pastures, and consequently decreased the total work time of farmers. Saddle horse breeding required labour, thus recruiting employees and volunteers. These stud farms also provided local services, so that they operated as small developmental poles in rural zones. Finally, most of these surveyed farms had been run for several years in their current operating system, which breeders wanted to sustain.

These first results must be completed to precise how equine production could impact positively on farm sustainability in each agricultural context and particularly in more cultivated regions. One other perspective to our work would be to assess the agro-ecological profit associated to maintenance of grasslands grazed by horses and cattle.

Acknowledgements

This work received financial support from the Institut Français du Cheval et de l'Equitation Paris, and from the Scientific Council of AgroSup Dijon, France.

References

Agreste Auvergne, 2012. Deux vaches nourrices pour une vache laitière sur les herbages auvergnats. Recensement agricole 109.

Agreste Franche-Comté, 2012. Les fondamentaux franc-comtois demeurent: un élevage bovin extensif herbager. Recensement agricole 177.

Agreste Basse-Normandie, 2012. La vache normande: une signature régionale. Recensement agricole 58.

Bigot, G., E. Perret and N. Turpin, 2010. L'élevage équin un atout pour la durabilité des territoires ruraux: cas de la région Auvergne. ASRDLF-AISRe. Septembre 20-22. Aoste, Italie.

Bigot, G., A. Célie, S. Deminguet, E. Perret, J. Pavie, and N. Turpin, 2011. Exploitation des prairies dans des élevages de chevaux de sport en Basse-Normandie. In: Bigot, G., W. Martin-Rosset, B. Morhain and C. Vial-Pion (eds.) L'utilisation des ressources prairiales et du territoire par le cheval. Fourrages 207: 231-240.

Bigot, G., G. Brétière, L. Viel, A. Célie, E. Perret and N. Turpin, 2012a. Farm grazing management with horses in two French grasslands regions: Normandy and Auvergne. In: Saastamoinen, M., M.J. Fradinho, A.S. Santos and N. Miraglia (eds.) Forages and grazing in horse nutrition. EAAP publication no. 132. Wageningen Academic Publishers, Wageningen, the Netherlands, pp 209-212.

Bigot, G., S. Mugnier, E. Perret, S. Ingrand and N. Turpin, 2012b. Le cheval et le pâturage des prairies dans les exploitations agricoles. Equ'idée 81: 29-31.

Bigot, G., G. Brétière, D. Micol and N. Turpin, 2013. Management of cattle and draught horse to maintain openness of landscapes in French Central Mountains. In: 17th Meeting of the FAO-CIHEAM Mountain Pasture Network Pastoralism and ecosystem conservation. June 5-7. Trivaro, Italy, pp. 72-75.

Fleurance, G., N. Edouard, C. Collas, P. Duncan, A. Farruggia, R. Baumont, T. Lecomte and B. Dumont, 2012. How do horses graze pastures and affect the diversity of grassland ecoysystems? In: Saastamoinen, M.T., M.J. Fradinho, A.S. Santos and N. Miraglia (eds.) Forages and grazing in horse nutrition. EAAP publication no. 132. Wageningen Academic Publishers, Wageningen, the Netherlands, pp 147-161.

Janssen, S. and M.K. Van Ittersum, 2007. Assessing farm innovations and responses to policies: a review of bio-economic farm models. Agricultural systems 94: 622-636.

Miraglia, N., 2012. Equids contribution to sustainable development in rural areas: a new challenge for the third millennium. In: Saastamoinen, M., M.J. Fradinho, A.S. Santos and N. Miraglia (eds.) Forages and grazing in horse nutrition. EAAP publication no. 132. Wageningen Academic Publishers, Wageningen, the Netherlands, pp 439-452.

Mugnier, S., G. Bigot, E. Perret, C. Gaillard and S. Ingrand, 2014. Dairy cow's pastures quality in Jura Mountains and Comté cheese area: maintenance with draught horses. In: Forage resources and ecosystem services provided by mountain and Mediterranean grasslands and rangelands. OPTIONS Méditerranéennes 109: 193-196.

Perrot, C., G. Barbin, N. Bossis, F. Champion, B. Morhain and E. Morin, 2013. L'élevage d'herbivores au recensement agricole 2010. Le dossier Economie de l'Elevage 440-441, 96 pp.

REFErences, 2011. Panorama économique de la filière. IFCE, le Pin-au-Haras, Paris, France, 241 pp.

REFErences, 2013. Annuaire écus 2013. IFCE, le Pin au Haras, Paris, France, 63 pp.

Rook, A.J., B. Dumont, J. Isselstein, K. Osoro, M.F. Wallis de Vries, G. Parente and J. Mills, 2004. Matching type of livestock to desired biodiversity outcomes in pastures – a review. Biological Conservation 119: 137-150.

Part 8.

The case of horse meat

The penultimate part of this book features what is, in effect, a World First. Although a controversial topic, the idea of using horses as food is returning from a not-so-distant past where horse milk and meat was a general part of the human diet. Like so much traditional ecological knowledge, it turns out that horse meat is nutritious and can be a welcome source of relatively low cost protein in some parts of the world. As one author claims, there is the additional issue of what to do with so many leisure horses, especially as they get older, with considerable controversy around euthanasia and the growing challenges of disposing of dead horses. And, again, as he claims, in certain societies, especially in the east and south of Europe, the practices of eating horse meat have never gone away, meaning that a significant proportion of European horses exist because of their role in the food chain, and if that role is excised, then the breeds and populations face disappearing. Once again, the issue of economic imperative is raised.

Seeing a cautious growth of the consumption of horse meat, W. Martin-Rosset has created what may be the first complete guide to the production of horses for human consumption. His chapter discusses everything from breeding and weaning, to winter feeding, to slaughter and handling of the meat products. Here you will find a basic guide to everything a farmer needs to know and consider when thinking about moving into this market. Martin-Rosset's chapter is the culmination of a long career as an equine scientist, and as past-president of the EAAP Horse Commission, he has been particularly well placed to both review the research done on the topic, and to synthesize it into the guidance and advice here. We believe that this chapter will remain a 'go-to' reference for those moving into the field for a long time to come.

17. Promoting slaughtering of horses and consumption of horsemeat – ethical horse keeping and meat production

M.T. Saastamoinen

MTT The Natural Resources Institute Finland (Luke), Equines, Opistontie 10 a 1, 32100 Ypäjä, Finland; markku.saastamoinen@mtt.fi

Abstract

It can be calculated, based on data from Sweden and Finland that about 5% of hobby and athletic horses die or put down every year mainly due to illness or old age. Consequently, it can be supposed that about 300,000 horses in Europe die yearly, not included those raised for meat production. Only part of those 300,000 former hobby and athletic horses are slaughtered in their home countries and part is transported to be slaughtered in other countries. Those not slaughtered are put down by other ways and usually buried. Both long transportation and burying the bodies is ethically and ecologically unsustainable. Encouraging horse owners to sell their horses to slaughterhouses and customers to use more horsemeat may be a good welfare-supporting issue, at the end of a horse's life. Increasing national and local horsemeat production will promote establishing small scale slaughterhouses in rural and peri-urban areas. Concerning specialized quality horse meat production based on native and local breeds, it may be one way to support these breeds and, thus to diversity of horse populations in cases when the breeds are endangered. In Finland, two projects promoting horsemeat consumption and encouraging horse owners to sell their horses to slaughterhouses have been carried out between 1998 and 2009. In Sweden, a corresponding project was carried out in 2012. The main problems that horse slaughtering raised in those projects were the safety of the raw meat material, the low slaughter price, long transportation distances, high slaughtering cost and unwillingness of owners to sell horses to food production because of uncertainty of the handling of the horse during slaughtering. This paper discusses the results of the projects and other euthanasia alternatives as well as ethical and environmental issues concerning putting down of horses after their active career.

Keywords: horse, welfare, native horse breeds, horsemeat scandal

Introduction

The number of horses in Europe is approximately 6,000,000 (EHN, 2013). Most of the horses today are used for pleasure and recreation purposes, including riding, galloping and trotting sports. Other uses of coldblooded horses are e.g. farming and logging, but the farm-work uses are decreasing dramatically in those countries, even where it still exists (Kozak, 2013; Maijala, 1999). In some countries coldblooded horses are bred for meat – and in smaller extent for milk – production (e.g. Mantovani *et al.*, 2005; Saastamoinen and Mäenpää, 2005).

Should we be happy with the increasing number of horses? We know that the useful life of the horse is limited, but what happens then? There is a time when horse owners have to part with their horses for some reason (old age, severe medical illness/injury, handling problems, poor economy). It can be calculated, based on registration data from Sweden (HNS, 2103) and Finland (Suomen Hippos, 2014) that about 5% of hobby and athletic horses die or put down every year mainly due to illness or old age. Consequently, it can be supposed that in Europe 300,000 athletic and hobby horses die every year. In Finland about 50% (Suomen Hippos, 2014) and in Sweden about 20% (HNS, 2013) of those horses are slaughtered for meat production in their home country, which can be considered generally rather low. It is obvious that in other countries the proportion of slaughtered horse is within this variation. Regarding the rest of dead horses, the bodies are destroyed or buried, but live horses are also transported to other countries to be slaughtered there (HSI, 2013). In many cases, slaughtering of athletic and hobby horses for meat is not possible because of medication.

Consequently, the owners have to consider how to get rid of the horse ethically and ecologically, i.e. in an acceptable way. We have to keep in mind that the end of the horse's life is important, as well as the road to that end. The possible choices are slaughtering or euthanasia at home stable or horse clinic. When putting the horse down in the latter way, the problem is what to do with the dead body. The alternatives are burial (where allowed), cremation, rendering, destruction – all of these causing high costs. Further, are these acceptable by the society; are they sustainable, ecological and ethical? For example, burying the dead body, legally or illegally, causes a risk to the environment if not done properly. In many areas there is not the area of land available for burying large animals. Further, high costs may lead to abandoning of horses –there are lot of abandoned horses for example in the UK and Ireland, and this is becoming a big welfare question.

Slaughtering of horses can be considered as an ethical way to put down horses from the point of view of animal welfare, environmental impacts and human food production. It is also an economical way to put down a horse. In addition, it is one way to support endangered native horse breeds when those, usually heavy breeds, are kept for meat production. In Finland, two projects promoting horsemeat consumption and encouraging horse owners to sell their horses to slaughterhouses have been carried out between 1998 and 2009. In Sweden a similar kind project was carried out in 2012.

Why slaughtering?

Slaughtering for human consumption is ethical, economical and sustainable from the point of view of animal welfare, environmental impacts and human food production. First, if the horsemeat is not used for human consumption, it can be considered as a waste of food. In the same time, the demand for animal protein and consumption is constantly increasing all the time around the world (FAO, 2009). Second, horsemeat has many advantages compared to other animal's meat: its nutritional properties are good (Martin-Rosset, 2001), and the ecological footprint is relatively low (Seppälä, 2013). Horses participate in the maintenance of the biodiversity, and moreover, they could sustain the development of local areas (Miraglia et al., 2006).

The ecological foot print of horse is only 1.6 ton greenhouse gases per year, which is 10% of that of people (Seppälä, 2013). Compared to beef cattle the annual greenhouse gas emissions of horses are about one quarter (Regina et al., 2014). If the meat from the body of the horse is utilised as human food, the ecological footprint of the horse can be even smaller.

In many countries there is an increasing demand of horsemeat from customers. However, the domestic supply of horses for slaughter is usually small and below the consumption, although people prefer domestic meat (Cressent and Jez, 2013; HNS, 2013; Laatuketju, 2010; Jaskari et al., 2015). Thus, horsemeat is imported into many countries where it is consumed more than produced (HNS, 2013; Jez et al. 2013a; Laatuketju, 2010; Martuzzi et al., 2001; Poncet, 2007) and there are several exporting countries selling low-price horsemeat (e.g. HSI, 2013; Martuzzi et al., 2001). For example, more than 30,000-50,000 live horses are imported to Italy from other European countries (HSI, 2013, 2014) yearly. The long transportation through Europe have, consequently, become a big and well known problem, causing severe injuries to the horses (World Horse Welfare, 2008). Domestic slaughtering results in shorter and safer transportation and better animal welfare, due to proper control of transportation and slaughter facilities.

Another benefit is that horses utilize domestic feeds and when kept on pastures, they use local land areas positively influencing biodiversity and landscape health (Miraglia et al., 2006). National and local horse meat production and slaughtering would promote establishing and economy of small scale slaughterhouses in rural and periurban areas thus giving economic and employment benefits to local communities.

In those countries that have special horsemeat (or foalmeat) production, cold-blooded horses are used for this purpose (e.g. Baudoin, 1990; Mantovani *et al.*, 2005; Martin-Rosset, 2001; Martuzzi *et al.*, 2001; Poncet, 2007). Those and other corresponding breeds are usually historic native and national breeds originally bred for farm work, logging, forestry, etc. Today, many of those horses are endangered (Bodo *et al.*, 2005) because they are not needed any more in farm work. One way to conserve these breeds and thus maintain biodiversity is to use them in horsemeat production. There are possibilities to create local niche products and horsemeat brands, as has been done in France (Cressent and Jez, 2013). In a French study dealing with future scenarios, one scenario based on horses' public and socio-economic value resulted in conservation and branding of local horse breeds (Jez *et al.*, 2013b). The first European study dealing with consumer demand and willingness to pay for indigenous cattle product (Finncattle meat) show that 86% of the respondents would like to buy it and about a quarter of respondents would be willing to pay a higher price for it than for conventional meat (Tienhaara *et al.*, 2013).

What will happen if we don't have domestic slaughtering of horses? The U.S. is a good example of this: the U.S. banned horse slaughter in 2007 (GAO, 2011). After that, many animal welfare problems arose. Horse owners have fewer options for getting rid of horses they no longer want because of the high costs of feeding, caring and medication. This situation has led to a growing number of abandoned horses. It has also negatively affected the prices of lower-to-medium value horses, reducing them by 8 to 21% (GAO, 2011). The most important consequence, from the animal welfare point of view, is that horses are transported to Canada and Mexico to be slaughtered there. The horsemeat from Canada and Mexico is thus born and raised in the U.S. 140,000 live horses are exported to these countries each year from the U.S. Before being sent to slaughtering facilities, the horses may be sent to fattening feedlots in those countries (GAO, 2011). The same kind of problems can also be found in some European countries where people do not accept horse slaughtering or do not eat horsemeat. Because of this, the long transportation (to slaughterhouses or feedlots) of live animals is a big ethical and animal welfare question. This is why several states in the U.S. are seeking to reopen horse slaughter facilities again (GAO, 2011).

Are people eating horsemeat?

There are great cultural differences in how people respond to eating horsemeat. It is difficult to break these cultural norms and structures to get people eat horsemeat (Jaskari *et al.*, 2015). Some of the meanings attached to horsemeat consumption appeared to be rather similar to those related to consumption of pork and beef, e.g. mainly ethical (Jaskari *et al.*, 2015). People are often out of touch with the realities of slaughtering and animal production (Buscemi, 2014; Jaskari *et al.*, 2015).

In many societies, the horse has had an iconic role; its former importance has been as a work and transportation animal. It was also 'a war hero'. Today its value is as a racing, recreation or show animal. For many people, horses are 'companions or pet animals' and they don't want to eat them. The horse is still also a working animal in many countries (Bodo *et al.*, 2005; Maijala, 1999). However, according to the EU regulation the horse is a production or a farm animal, and, thus, a part of the food chain. More detailed discussion about the reasons why people are not eating horsemeat is presented by Jaskari *et al.* (2015). For example, healthy risks due to remnants of medical drugs are one reason to avoid eating horsemeat.

There are some countries and cultures, e.g. UK and Ireland, where people do not eat horsemeat at all. In many countries this is based on Christian traditions; the pope Gregorius III banned horsemeat eating in 732. Since the 15th century, eating horsemeat has been a taboo in Europe. This changed in the late 1800's because of the lack of food and for animal welfare reasons. Today, consumers in some countries or areas traditionally have a strong preference for horsemeat, e.g. in Italy and France

(Martin-Rosset, 2001; Martuzzi *et al.*, 2001). More detailed description about the history of human use of horsemeat is presented by Jez *et al.* (2013a).

Horse slaughtering and human consumption of horsemeat in EU

According to statistics, over 250,000 horses are slaughtered for human consumption in the EU each year (HSI, 2014). The numbers of equids (mainly horses) slaughtered for meat in different European countries are presented in Table 1. Nearly half of these animals are slaughtered in Italy, where the demand for horse meat increased dramatically due to BSE at the beginning of the 2000's leading to the situation where more than half of the meat supply comes from importation (Martuzzi *et al.*, 2001). The remainder are slaughtered primarily in Poland, Spain, France, Ireland, Romania, Belgium and Germany. In some countries, e.g. in Poland and Germany, the numbers of slaughtered horses have increased clearly after 2007. Lower levels of horse slaughter are recorded in the Czech Republic, Latvia, Lithuania, Luxemburg and other countries. Detailed data on horse meat trade in EU can be found in www.hsi.org.

It is mainly low-priced horses that are more likely to be bought for slaughter in the European countries. The number of slaughtered horses in Europe has increased due to the current economic downturn. For example, according to the news, it has doubled in Spain from year 2007 to at least 60,000 horses in 2013; however, the official number given by HSI (2014) is about 50,000 (Table 1).

The largest horsemeat consumer countries are Italy, France and Belgium. Table 2 shows the amount of horsemeat production in some EU countries. The average consumption of horsemeat in EU countries is about 0.4 kg/inhabitant/year, but in Italy it is about 1.3 kg (Martin-Rosset, 2001; Martuzzi *et al.*, 2001). Consumption has decreased in many countries after 1970's and 1980's (Cressent and Jez, 2013a; HSI, 2013; Martin-Rosset, 2001). On the contrary, in some countries the consumption has increased, however, due to having a low base rate of consumption, like in Finland and Sweden. In those countries the number of slaughtering facilities has also increased. In general, in the EU countries a large number of slaughter houses are registered to slaughter horses – in many countries they are spread quite equally (World Horse Welfare, 2008). These circumstances make it easy for the horse owners to get their horses slaughtered.

Poland, Spain and Romania are the main exporters of slaughter horses and horsemeat in the EU. Numbers from Romania are not available, but the 'horsemeat scandal' has recently shown that

Table 1. Horses slaughtered for meat production in 2007 and 2013 (HSI, 2012, 2014).[1]

Country	2007		2013	
	%	Heads	%	Heads
Italy	46.6	99,970	24.0	51,845
Poland	18.4	39,608	10.4	22,514
Spain	12.2	26,172	23.3	50,319
France	8.2	17,744	9.5	20,544
Belgium	4.7	10,149	4.0	8,734
Germany	4.5	9,704	4.9	10,613
Sweden	1.3	2,996	1.7	3,785
The Netherlands	1.2	2,656	2.2	4,700

[1] 2007 is the last year for which the most complete statistics on horse slaughter from EU member states are available (HSI, 2013). For 2013 data for slaughtered equids (HSI, 2014).

Table 2. Horsemeat production in some European countries.

Country	Tons per year	Calculated from
Italy	18,000	Farming in UK, 2009
Poland	16,000	Farming in UK, 2009
Romania	14,000	Farming in UK, 2009
France	7,500	Farming in UK, 2009
Sweden	1,360	HNS, 2013
Finland	500	Laatuketju, 2010

Romania is a big horsemeat producer. Germany, Denmark and The Netherlands are also exporting horses and horsemeat. The largest importers of horsemeat and live slaughter horses inside Europe are Italy, France and Belgium. For example, 30,000-50,000 live horses are imported yearly to Italy to be fattened and slaughtered there. Thus, long transportations are a problem also in Europe and have been under strict critic (World Horse Welfare, 2008). Regarding importing outside the EU, Belgium is a major importer of horsemeat. The main exporters of horsemeat to EU are Canada, Argentina, Brazil, Mexico and Uruguay. The EU requires lifetime medication records for all horses slaughtered in non-EU countries before accepting imports of horsemeat from those countries, to ensure the safety of the meat.

The 'Horsemeat scandal'

In the 2013 meat adulteration scandal, foods advertised as containing beef were found to contain undeclared horsemeat – as much as 100% of the meat content in some cases (Wikipedia, 2014). The issue came to light on 15 January 2013, when it was reported that horse DNA has been discovered in frozen beefburgers sold in several Irish and British supermarkets.

While not a direct food safety issue, the scandal revealed a major breakdown in the traceability of the food supply chain, and therefore some risk that harmful ingredients were included as well. Sport and hobby horses for instance could have entered the food supply chain, and some of them could have been treated with veterinary drugs which are banned in food animals. The scandal has since spread to 13 other European countries and European authorities have decided to find an EU-wide solution.

The horsemeat that was found in some food products came from a Romanian slaughterhouse. An inquiry by the French government showed that 'the meat had left Romania clearly and correctly labelled as horse. It was afterwards that it was relabelled as beef.' (BBC, 2013a,b,c). There were several dealers in various countries involved in the trade of the meat. After processing, the meat was sent to another company where the end products for sale were made. According to French media reports, one trader falsified documents regarding the meat. Some of the 'scandal horsemeat' found in products is thought to have originated from Poland (Wikipedia, 2014). Thus, the 'horsemeat scandal' was not caused by the horse sector or horse slaughtering but was a consequence of criminal action of meat dealers and traders in Europe.

What were the consequences of those happenings to horsemeat's image and consumption? The whole process with news and talks about horsemeat gave opportunities to show the positive aspects and properties of horsemeat, including its healthy nutrient content compared to that of other animals' meat. Human consumption and demand has increased in some countries, e.g. in Finland and Sweden. Especially, interest and demand for *domestic* horsemeat increased. However, some meat houses stopped using horse meat, but in these cases, horsemeat was only a minimal part in their products.

Some slaughterhouses also stopped slaughtering horses to minimizing the risk of contamination with horsemeat.

Projects promoting horse slaughtering and human consumption

It is reasonable to promote horse slaughtering and horsemeat consumption from the ethical and animal welfare point of view. There are also other reasons, as described above. From many points of view, reasons for decrease in slaughtering are not sustainable. It is important that 'the horse lovers' too can understand the welfare and ethical issues of slaughtering.

To promote human horsemeat consumption and encourage horse owners to send their horses to slaughterhouses when the horse's life is in the end, some national projects have been undertaken. These projects have been focused on horse owners, slaughterhouses, meat industry and consumers. The main emphasis has been on animal welfare, economy of horse keeping and quality and safety of horsemeat. One purpose has also been to increase the price of slaughtered horses – horsemeat lost its price advantage in 1990's, which is also one reason for decreased slaughter levels (Jez *et al.*, 2013a). Generally, the prices of slaughtered horses are very low, e.g. around 50 cents/kg (HNS, 2013; Laatuketju, 2010), but also some much higher prices (4-5€/kg) are paid for foalmeat in some countries (Poncet, 2007). Thus, the aim was to increase of the supplying of horses for slaughter and human consumption of domestic horse meat.

Two special projects or campaigns in Finland (1998, 2009) and one in Sweden (2012) have been carried out for those purposes, in collaboration with horse breeding and owner associations, producers' associations and Ministry of Agriculture. However, we can say that 'the first campaign' in Finland was the establishment of an association for animal welfare in 1901, one of its main aims being to promote human use of horsemeat in that time.

In those programmes (HNS, 2013; Laatuketju, 2010), horse owners were informed about the advantages and especially about the ethics of slaughtering. Slaughter- and meat houses were helped to get horses for slaughtering in such amounts that the slaughtering is profitable. Customers were encouraged to consume horsemeat in various forms and slaughterhouses were helped in marketing of the meat and meat products. The bottlenecks of slaughtering, marketing and consumption of horsemeat were studied.

The results and conclusions of those projects show that there are generally many difficulties in trying to increase the slaughtering of horses, mainly because of the long distances to horse slaughterhouses (HNS, 2013; Laatuketju, 2010). The main problems were the low price of a slaughtered horse, and at the same time the high costs for transportation. It is also evident that all horses can't be slaughtered because of medication and insufficient medication recording. From the point of view of the slaughterhouses one problem is uneven supply of horses for slaughter. The main problems included insecurity of the safety of the meat raw material (identification, medication), and unwillingness of owners to sell horses to food production because of uncertainty of the handling of the horse during slaughtering. It is also more expensive to handle horses in the slaughter facilities than carcass of other species, mainly because of the fact that only 12% of the horse carcass is valuable meat in the markets. Special attention should be paid to the transportation length and circumstances (temperature, humidity, space, risk of injuries, etc.) in the vehicles used for slaughter transportations.

In France, horsemeat industry stakeholders organized in 2002 to support human horsemeat consumption. This organization included breeding organisations of nine draft horse breeds. It developed local meat brands and gave advices to producers (Cressent and Jez, 2013). According to Jaskari *et al.* (2015) horsemeat can be seen as a culinary delicacy. There is potential for new product development. Further, health-related values of the horsemeat may also offer new market opportunities (Jaskari *et al.*, 2015).

Conclusions

Horse has an important role in the food chain. Many people are interested in eating horsemeat, and customers want to buy *domestic* horsemeat and horsemeat products. The interest for horsemeat has even increased during 'the horsemeat scandal', and more supply of both horses to slaughtering and meat in the shops and restaurants are needed. Promoting and encouraging horse owners to send their horses to domestic slaughterhouses, is important. Slaughtering is an ethical, economic and sustainable way to put down horses. Human consumption of horsemeat has economic and employment benefits for local communities. In addition, horsemeat production may be important to protect old native heavy breeds. More information is, however, still needed regarding slaughter as an alternative to put down a horse. Proper use of horse passport, documentation of all medications and identification of the horse are key factors to guarantee food safety, and confidence and to meet the growing demand for horsemeat by meat industry and customers. The results of the projects presented here are in accordance with the conclusions by Jaskari *et al.* (2015) who concluded that in marketing of horsemeat cultural and other reasons that influence horsemeat consumption should be considered.

References

Baudoin, N., 1990. Present state of horse breeding in France. In: Rossier, E. (ed.) Horse breeding in France. EAAP Publication 53. C.E.R.E.O.P.A., France, 106p.

BBC, 2013a. Horsemeat found in beefburgers on sale in UK and Ireland. BBC News 15 January 2013. Available at: www.bbc.co.uk/newsa/world-europe-21034942.

BBC, 2013b. 'Horsemeat beefburgers' investigated in UK and Ireland. BBC News 16 January 2013. Available at: www.bbc.co.uk/news/worl-europe-21038521.

BBC, 2013c. Horse meat scandal: France blames processor Spanghero. BBC News. 14 February 2013. Available at: www.bbc.com/news/world-21464052.

Bodo, I., A. Lawrence and B. Langlois (eds.), 2005. Conservation genetics of endangered horse breeds. EAAP Publication 116. Wageningen Academic Publishers, Wageningen, the Netherlands, 187 pp.

Buscemi, D., 2014. From killing cows to culturing meat. British Food Journal 116: 952-964.

Cressent, M. and C. Jez, 2013. The French horse industry at present. Advances in Animal Biosciences 4(s2): 54-56.

EHN, 2013. The horse in Europe. European Horse Network, 4p. Available at: www.europeanhorsenetwok.eu.

FAO, 2009. The state of food and agriculture: Livestock in the balance. Available at: www.fao.org/docrep/012/i0680e00.htm.

Farming in UK, 2009. Farming in UK, January 17 (2009). www.farminguk.com.

GAO, 2011. Horse welfare. Action needed to address unintended consequences from cessation of domestic slaughter. U.S. Government Accountability Office, Report to Congressional Committees, 62p.

HNS, 2013. Projektet hästliv. 23 p. Available at: www.nshorse.se.

HSI, 2013, 2014. Humane Society International. Facts and figures on the EU horse meat trade. 8p. Available at: www.hsi.org.

Jaskari, M.M, H. Leipämaa-Leskinen and H. Syrjälä, 2015. Revealing the paradoxes of horsemeat – The challenges of marketing horsemeat in Finland. Nordic Journal Business 64: 86-102.

Jez, C., B. Coudurier, M. Cressent and F. Mea, 2013a. Factors driving change in the French horse industry to 2030. Advances in Animal Biosciences 4(s2): 66-105.

Jez, C., B. Coudurier, M. Cressent and F. Mea, 2013b. Scenarios. Advances in Animal Biosciences 4(s2): 106-115.

Kozak, M.W., 2013. Equestrian tourism in Poland: status, opportunities, and barriers affecting local development. Folia Turistica 28(2): 205-226.

Laatuketju, 2010. Hevosenlihan markkinoiden tehostaminen. Loppuraportti. Maa ja metsätalousministeriö ja Suomen Hippos, 18p.

Maijala, K. (ed.), 1999. Use of horses in forestry and agriculture – breeding of working horses. Proceedings of the International Seminar on Working Horses. Special publication of the Finnish working horse society, Helsinki 1999, 128 pp.

Mantovani, R., G. Pigozzi and G. Bittante, 2005. The Italian Heavy Draught Horse breed: origin, breeding program, efficiency of selection scheme and inbreeding. In: Bodo, I., A. Lawrence and B. Langlois (eds.) 2005. Conservation genetics of endangered horse breeds. EAAP Publication 116. Wageningen Academic Publishers, Wageningen, the Netherlands, pp. 155-162.

Martin-Rosset, W., 2001. Horse meat production and characteristics. In: Book of abstracts of the 52nd Annual Meeting of the European Association for Animal Production. Wageningen Academic Publishers, Wageningen, the Netherlands, p. 322.

Martuzzi, F., A.L. Catalano and C. Sussai. 2001. Characteristics of horse meat consumption and production in Italy, 11 pp. Available at: www.unipur.it/arpa/facvet/annali/2001/martuzzi.pdf.

Miraglia, N., D. Burger, M. Kapron, J. Flanagan, B Langlois and W. Martin-Rosset, 2006. Local animal resources and products in sustainable development: role and potential of equids. In: Rubino, R., L. Sepe and A. Gibon (eds.) Livestock farming systems. Product quality based on local resources leading to improved sustainability. EAAP Publication 118. Wageningen Academic Publishers, Wageningen, the Netherlands, pp. 217-233.

Poncet, P.-A., 2007. Berich der Arbeitsgruppe Pferdebranche, 2007. Wirtschafts-, Gesellschafts- und Umweltpolitische Bedeutung des Pferdes in der Schweiz. Bericht der Arbeitsgruppe Pferdebranche, Avences, Switzerland, 160 pp.

Regina, K., T. Lehtonen and S. Ahvenjärvi, 2014. Agricultural greenhouse gas emissions and their mitigation. Available at: http://www.mtt.fi/mttraportti/pdf/mttraportti127.pdf.

Saastamoinen, M.T. and M. Mäenpää, 2005. Rare horse breeds in Northern Europe. In: Bodo, I., A. Lawrence and B. Langlois (eds.) Conservation genetics of endangered horse breeds. EAAP Publication 116. Wageningen Academic Publishers, Wageningen, the Netherlands, pp. 129-136.

Seppälä, J., 2013. Lemmikkieläinten ja hevosten hiilijalanjälki. Turun Sanomat 24.3.2013.

Suomen Hippos, 2014. www.hippos.fi.

Tienhaara, A., H. Ahtiainen and E. Pouta, 2013. Consumers as conserves – Could consumers' interest in a specialty product help to preserve endangered Finncattle? Agroecology and Sustainable Food Systems 37: 1017-1039.

Wikipedia, 2014. 2013 meat adulteration scandal. Available at: https://en.wikipedia.org/wiki/2013_meat_adulteration_scandal.

World Horse Welfare, 2008. Dossier of evidence. Recommendations for amendments to EU Council Regulation (EC) No 1/2005. World Horse Welfare, United Kingdom, 55 pp. Available at: http://tinyurl.com/pxwnqbk.

18. Horse meat production and characteristics: a review

W. Martin-Rosset[1]* and C. Trillaud-Geyl[2]
[1]*INRA, Center of Research of Clermont-Ferrand / Theix, 63122 Saint-Genès-Champanelle, France;*
william.martinrosset@gmail.com
[2] *IFCE, French Institute for Horse and Riding, Terrefort, BP 20, 49411 Saumur, France*

Abstract

Several countries are consuming horse meat in Europe. Horse meat is supplied either with draft breed colts raised by herds of mares managed in extensive conditions or with riding hoses culled in Europe or imported from overseas countries where there is no tradition for horse meat consumption. Husbandry and production systems using draft breeds have been extensively studied. Herds of mare can be managed at low cost on extensive grassland in mountainous and hilly areas with good productivity. Various production systems for fattening colts and their carcass and meat characteristics are defined according to age, breed, sex and diet. Carcass and meat characteristics of culled riding horses are described too. Consumer's preferences, nutritional and hygienic characteristics, and technological horse meat transformation for human consumption are well established. Prospect for horse meat production and/or supply in different European countries are discussed.

Keywords: horse, meat, production, characteristics, technology

Introduction

Horse meat is consumed in several European countries (Belgium, France, Italy, Slovenia, Spain and Switzerland) since long time, and likely in Russia. Horse meat is supplied with riding horses and draft horses culled and/or fattened respectively. Culled horses are either locally supplied or extensively imported from other European countries and America (Canada, United States and Argentina) where there is no tradition for horse meat consumption. Consumption needs relies mainly on culling the riding horses. But in few European countries draft horses are raised and then fattened for meat consumption using their large population of draft breeds that are no more used for agricultural work. France is highly concerned regarding its large draft horse breeds population and the high proportion of horse meat consumption supplied by importation of culled light breed horses. Hence, horse meat production and characteristics were studied in the scope of a large program supported by French government in the 70's and 80's. Horse meat supplied by culled light breed horses has been also studied to a lesser extent in France and in Switzerland. Characteristics and technology of horse meat have been very well studied in Italy and in Russia.

Husbandry and production systems with draft breeds: experimental data

There are still 80,000 draught horses in France which become available for meat production (ECUS, 2012). Mares are found either in the breed cradle areas: grass or arable zones north of the Loire, where the number of animals is dropping rapidly and is very much spread or mainly in hilly and mountains areas where variably sized herds have been constituted for horse meat production.

Before 1965, internal production of horse meat came mainly from slaughter of last working horses. More recently, fattening of colts for meat has developed in arable areas, the fringes of extensive areas and, to a lesser extent, in grasslands zones.

Long term research in the field was carried out from the year 1970 to 1984 at INRA and National Stud (so called today: Institut Français du Cheval et de l'Equitation, IFCE; National Institute for Horse and Riding in English).

Management of herds of brood mares

In North-East grassland zones, particularly with extensive systems (Lorraine, Franche Comté, and Limousin) herds of mares are raised with beef or dairy herds (Bigot *et al.*, 2011; Morhain, 2011).

Mares generally give birth at the beginning of April, before being turn out to grass, and are weaned during the second half of October, after 190 to 210 days lactation. The foals weight 350 to 400 kg at weaning. Productivity measured as number of foals weaned per 100 mares covered is satisfactory: 70 to 75%. Winter feeding lasts 110 to 120 days, for grass reserves left by ruminants at the end of the summer season are rapidly used during the autumn. In winter mares are generally fed hay (1,500 kg) and 120 kg concentrates barley or maize harvested on the farm and complemented, in the best situations, by meal. Mares may also be fed straw free choice (900 kg over the winter), 1st cut hay or regrowth hay fed in limited quantities (500 kg) complemented with a concentrate (350 kg). They may be quite severely underfed since lactation only occurs during grazing, and underfeeding only during pregnancy, which lasts 120 to 150 days. However, this depends also on the possibilities of stocking rate (horses and ruminants) and summer grass production. Mare herds graze both areas not cut for hay or of poor quality, which they help to improve, and also in autumn the roughs left by ruminants during the summer. Pasture represents 70 to 80% of total annual needs. In all cases horses and ruminants are complementary on the farm, with few problems, except perhaps in spring, when a minimum number of good grazing are necessary for nursing mares (1st month of lactation).

This system makes use of poor grassland and maximises production per ha of grassland while reducing the labour necessary for cutting roughs. However it is relatively fragile, especially in the case of summer drought but possibility of grass or cereal culture, use of by-products such as straw, maize cobs, etc, or even maize or grass silage (35 and 50% DM respectively) give greater flexibility than with mountain herds.

Nevertheless, the grassland available for meat production is found in hilly and mountain areas (South-west: Pyrenees, Centre: Massif-Central, East: Vosges and Franche-Comté) and it is in these areas that there is the greatest development of herds of nursing mares. The results obtained in experimental condition (INRA Research Center of Research at Theix, in the North of Massif-Central: Auvergne, at 800 to 1,500 m altitude) exhibit the interesting possibilities (Figure 1).

The mares foal generally during the first week in April, a month before being put out to grass. They generally dry at the beginning of October, after 180-190 days lactation. The foal weights about 330 to 340 kg at weaning. Mares receive costly feeds in limited quantities (1,300 kg forage and 150 kg concentrate/mare for the winter) only during the last two months of pregnancy and the first month of lactation, at the end of winter. Mares adapt satisfactorily to underfeeding at the end of autumn after weaning, and during the winter, for 150 to 200 days/year. Underfeeding can be quite severe during the last 100 last days if the animals are in good condition (body condition score: 3.5 according to scoring scale 0-5; INRA *et al.*, 1997) in the autumn (780 to 800 kg live weight) and have abundant grass in the spring.

Grazing represents 80% of the animals' total annual needs, covering 9 months of the year. Mare herds use poorly productive areas which give 60 to 70% of animal feeds and almost all grazing. This system combines maximal use of several types of pasture, successively, which the horses maintain and improve (Martin-Rosset *et al.*, 1985b). During the spring, mares are turn out to grass, generally on poor quality, which cannot be cut for hay, or graze after dairy or beef cows, the better quality

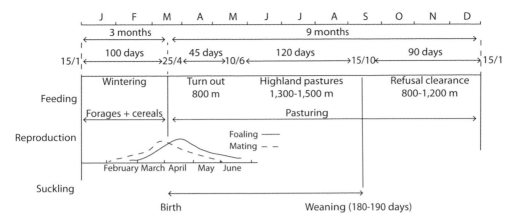

Figure 1. General diagram of brood mare management of draft breeds in France: mountainous areas (Martin-Rosset et al., 1985b).

pastures. From June onwards, the mares use degraded highland at 1,200 to 1,500 m altitude. These zones are either grazed directly by horses or after a first grazing by beef cows. At the end of the summer, in the autumn or the beginning of winter, when beef cattle are stabled, the horses clean up pastures close to the farm or communal grassland at 800 rn altitude, of the roughs left by ruminants during the summer.

This system is relatively efficient and economic, but requires a complementary area of grassland in the spring during the first weeks of lactation and the period of covering. The number of foals weaned 70-80% (Martin-Rosset *et al.,* 1985b) and their weight at weaning are satisfactory. 80 to 90% of the mares are fertilised. However the system is fragile and depends on annual pasture conditions.

Production and fattening of colts for meat

Various production systems have been defined (Figure 2). Research at INRA since 1972 has provided information concerning the growth of colts with good quality flesh and its variation with age, breed, sex and diet (Martin-Rosset *et al.,* 1985a). This work has defined the technical bases of various production systems that use through feeding or grass (Martin-Rosset *et al.,* 1985b), according to the type of animal, the nature of feeds available and the region. These are summarised in Table 1.

Colt production indoor

A colt weaned at 6-7 months is fattened indoor up to 10 to 15 months according to its weight at weaning and the energy concentration of the diet. lt may have a continuous growth of 1,000 to 1,400 g, close to the genetic potential (Figure 2). It is slaughtered at a live weight varying from 450 to 500 kg, according to breed (Martin-Rosset *et al.,* 1980). It produces a carcass of 270 to 300 kg (Table 2). The meat is tender but pale in colour (Figure 3).

The colt is fed a diet of high energy concentration composed of forages of high nutritive value and including 35 to 60% concentrate; based on cereals and soybean meal. The forage is fed free choice (Table 1). This, no doubt, represents one of the most efficient systems from a technical point of view of transforming feeds into meat, but it is very dependent on nutritive values and quality of feeds (Agabriel *et al.,* 1982; Martin-Rosset *et al.,* 1985b). It requires limited work and can be used in parallel, in the same building as bulls. It provides a rapid return for capital outlay. It is found in

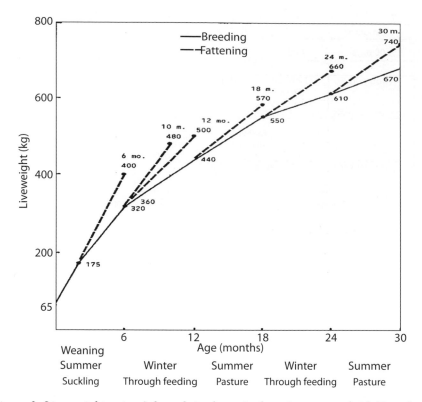

Figure 2. Live weight gain of the colt in the main fattening systems with French draft breeds. (Martin-Rosset et al., 1985b).

regions producing cereals and forages such as maize silage (Parisian region, central France, Brittany). However it is used increasingly on the forages of extensive zones where the mares are to be found.

Production of colts at grass

Pasture grass is, no doubt, still the least costly feed for horse meat production. Work from INRA and IFCE (Martin-Rosset *et al.* 1985b), indicates that it is possible, with horses to make use of intensive grassland to produce and fatten colts slaughtered at 18, 24 or 30 months of age (Figure 2).

Production of 18 months old colts

The colts are not raised to obtain, immediately after weaning, a maximum growth rate. During the winter following weaning, they have a moderate growth rate, followed by a high and compensating growth rate at pasture. They are slaughtered either at 18 months at the end of the grazing season, with a live weight of 550 to 580 kg and a carcass weight of 330 to 350 kg or at 22-24 months after finishing indoor during a second winter, with a carcass weight of 370-400 kg (Figure 2; Table 2). However, there are variations according to breeding and sex and also to grazing conditions. An 18 month colt must be produced from a foal with a good growth rate during nursing (1,400 to 1,700 g/day) and thus weaned at a minimum live weight of 330 kg. During the winter following weaning the animals should have a mode rate growth rate of 600-700 g/day in order to make best use of compensating growth at pasture (900-1,100 g). The colts are not castrated, for although this does

Table 1. Main production systems of horse meat from draft breeds in France (Martin-Rosset et al., 1985b).

Slaughter age	Winter feeding	Systems	Area of production
6-7 months	Milk + grass + concentrate	Heavy foal	Forage production
10-15 months	High quality forages (given free choice): early harvested hay maize silage ≥30% DM Concentrates: (35-60% of diet)	Intensive	Forages and grains production feedlots
18 months[1] 24 months[2]	Medium quality forage (given free choice) late harvested hay maize silage 25% DM grass silage 25% DM Good quality forage (given limited quantity): maize silage ≥30% DM grass silage prewilted >30%DM Concentrates: (10-20% of diet)	Semi-intensive	Forages production
30 months[2]	Poor quality forages (given free choice) late harvested hay Farming by products (given free choice): straw Concentrate: (5-10% of diet)	Extensive	Forages production or marginal lands

[1] Finishing with cereals on pasture.
[2] Finished in a feedlot.

Table 2. Slaughter characteristics of colts of French draft breeds (Martin-Rosset et al., 1980).

Slaughter age (months)	Body weight (kg)	Hot carcass (kg)	True yield weight (%)	Internal thoracic fat[1] (kg)	Carcass composition (%)		
					Muscles	Bone	Fat
6-7	380-420	220-240	67-69	3.0-4.0	68-69	17-18	9-11
12	470-500	290-310	70-71	4.0-5.0	70-71	16-17	8-10
18	550-580	330-350	68-69	2.5-3.5	69-72	16-17	9-12
24	630-660	370-400	68-70	5,0-7.0	68-69	14-15	13-14
30	740-780	430-460	70-71	8.0-10.0	68-69	14-15	13-14

[1] Internal thoracic fat: fat that adheres to internal side of sternum that allows fattening state to be evaluated accurately.

not reduce potential growth, it does not significantly help fattening and remains costly. In winter, the colts receive good quality forages such as maize or grass silage (35% DM) fed free choice with 5 to 15% concentrate, or less high quality forage, harvested late, with 15 to 20% concentrate (Table 1).

In summer they can use, in rotation, intensive pastures (mixture of ryegrass, bluegrass and fescue) receiving 100 units N/ha with a stocking rate of 2.5 animals/ha. 3 kg of ground maize must be

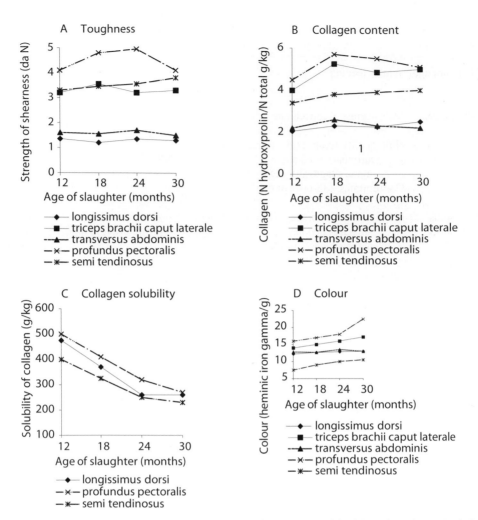

Figure 3. *Physicochemical characteristics evolution of horse meat of draft breeds with age (Robelin et al., 1984).*

distributed per day per animal during the last two months of grazing (15/8-15/10) if the animals are to be finished. Meat production from colts fed alone on grass is 350 to 450 kg/ha. However, wide variations occur according to grassland productivity (Table 3).

22-24 months colts

If the foals are too light at weaning less than 330 kg, their fattening must be completed indoor during a second winter, between 18 and 22-24 months, with a diet based on high value forage and 25 p.100 concentrates (Table 1). The colts are castrated at 18 months. The duration of finishing and the optimal level of winter feed depend on the previous growth rate and the breed. They remain to be determined.

The meat produced at 22-24 months is less tender but darker red than at 12 months (Robelin *et al.*, 1984).

Table 3. Grazing fertilized pastures by colts of draft breeds and beefs cattle (Martin-Rosset and Trillaud-Geyl, 2011; Martin-Rosset et al., 1985b).

Treatment	10 colts + 10 beefs	5 colts + 15 beefs
A Grazing		
Dates	25/4 to 3/11	
Duration (d)	192	
Total surface grazed (ha)	7.0	7.0
Proportion of horses in total body weight	48.9	33.2
Average stocking range (Al/ha)		
Horses[1]	1.47	0.74
Cattle[2]	1.21	1.89
Live body weight of horses at turning out to pasture(kg)	420.0	423.0
Live body weight of horses at end of pasture kg)	599.0	632.0
Average daily gain (g/d)		
Horses	977	1 148
Cattle	707	704
Output/ha (kg)		
Horses	260	154
Cattle	204	313
Total	464	467
Hay cut		
Date	25/4 to 2/6	
Surface (ha)	2.0	2.2
Dry matter (t/ha)	4.4	4.8
B Grazing		
Dates	5/4 to 3/10	
Duration (d)	181	
Proportion of horses in total body weight	59.9	31.41
Average stocking range (Al/ha)		
Horses[3]	1.52	0.77
Cattle[4]	1.46	2.20
Live body weight of horses at turning out to pasture(kg)	653.1	656.5
Live body weight of horses at end of pasture(kg)	752.7	793.4
Average daily gain (g/d)		
Horses	579	791
Cattle	793	1,009
Output/ha (kg)		
Horses	155	113
Cattle	242	360
Total	397	473
Hay cut	none	

[1] Colts of one year.

[2] Beefs of one year of 412 kg live body weight; beefs of two years of 525 kg live body weight, at turning out to pasture.

[3] Colts of two years of age.

[4] Beef cattle of 412 kg live body weight; beef cattle of 518 kg live body weight, at turning out to pasture.

These type of animals can be produced in grassland zones, where intensified pasture areas are available and where good pasture conditions can be secured. The animals may be fed during the winter with feeds harvested on the farm.

30 months colts production

This concerns only colts castrated at 18 months, but can be extended to fillies covered at 2 years but not fertilized or females slaughtered at 40 months in the absence of a second fertilisation after a first foaling.

In most cases, slaughter age is 30 months, particularly for colts with a live weight exceeding 700 kg, providing a carcass weighing 430 to 460 kg (Figure 2; Table 2).

The feeding level must be limited during the winter to obtain low cost maintenance and a slow growth rate: 500-700 g during the first winter and 260-300 g during the second winter, such that maximum use is made of grass, especially in the spring, through a high growth rate: 700 to 800 g/day during the first summer, 600 to 800 g during the second summer.

This type of animal makes good use of grass, due to a high intake capacity. Its production cost may be lower than that of 12 month old colts since it is slaughtered at a much higher live weight.

In our trials, live weight gain per ha varies from 400 to 600 kg. 30 months colts may be finished with grazing and without cereals, for the percentage of fatty tissue in the body increases strongly with weight (Martin-Rosset *et al.,* 1983). The meat produced is tenderer than at 24 months but less than at 12 months. It has the same red colour as that from adult horses.

This production system is not yet widely used, because it raises the problem of capital immobilisation. In addition, for production of this type of animal, the meat must be sold at a higher price than that from younger animals, due to its quality and the fact that the farmer requires greater technicity than in the case of intensive systems. Nevertheless, such animals can be produced in conjunction with cattle, either in grassland areas or in highland zones where pastures are of low value and ruminants do not utilize to a maximum their potentialities.

Mixed grazing

Horses may be associated with ruminants to make maximum use of grass production and to control the species in the grass sward, taking into account the specific effects of the feeding behaviour of both animals. Horse-cattle association: in trials described earlier, 1 to 2 year old colts may be associated with 1 to 2 years old steers. Animal behaviour is satisfactory. The horses and cattle grazed the same territory in separate groups. All the pasture was grazed such that it was not necessary to cut roughs. Meat production varied from 400 to 470 kg/ha when the horses were associated with cattle in the respective proportions 1/1 and 1/3 (Table 3A and 3B) but horses 'average daily gains decrease as proportions of horses increase.

The interest of mixed grazing remains a quantitative measurement. It is difficult to separate, among the data available, the effect of the combination of species and that of stocking rate on the variation in individual performance and production per ha. In contrast, it appears that, in the case of intensive natural grassland, exploited by horses associated with cattle, horses must not exceed 30% of the total stock. The percentage should correspond approximately with the proportion of grazing refused by the cattle.

Carcass characteristics

Draft horses

Variation of carcass characteristics: experimental data

The characteristics of carcasses and meat were stated between 12 to 30 months in long term experiments carried out in France at INRA and National Stud (IFCE) with young horses: 50% males – 50% females provided by the 5 main French draft breeds; the males were castrated at 18 months (Boccard, 1975; Martin-Rosset and Jussiaux, 1977; Martin-Rosset *et al.,* 1980, 1983; Robelin *et al.,* 1984).

Carcass weight and true yield

Live body weight increases from 483 to 735 kg between 12 to 30 months (Figure 2). As digestive content accounts for 10 to 13% of live body weight, empty body weight rises from 440 to 622 kg. Hot carcass weight ranges from 313 kg to 442 kg (Table 4). True carcass yield (e.g. the ratio: hot carcass weight / empty body weight) is on average 70% (69.3 to 71.2%) which is quite similar to what is observed for beef cattle.

Tissues composition of the carcass

Muscle proportion is on average 70 p100 whatever the age of animals (Table 4). The ratio muscle/ bone increases from 4.5 at 12 months to 4.8 at 30 months. Proportion of adipose tissues in carcass increases very quickly from 9 to 14%.These characteristics are very close to those observed in beef cattle (Charolais, Limousin: Robelin, 1978).

Adipose tissues in the carcass account for 88 p100 of total adipose tissues in the total empty body weight (86% at 12 months and 89% at 30 months). Subcutaneous and internal adipose tissues rise from 23 to 31 and 10 to 13% respectively (Table 5). This evolution is quite similar to that observed in beef cattle (Robelin, 1978). Subcutaneous and internal adipose tissues are faster, than that of intermuscular adipose tissue. As a result the absolute proportion of subcutaneous and internal adipose tissue is very high in horse compared to beef cattle and the proportion of intermuscular adipose tissues in horse is lower between 6 to 30 months of age than that observed in beef cattle: 47 vs 70% respectively. The low proportion of intermuscular adipose tissue in horse can be associated to a low lipid intramuscular content. And this can argue the dietetic value of horse meat.

The relative variation of the weight between the different muscular area is very low (Table 6). Muscle development reaches very quickly a maturity threshold, which is very different to what is observed in beef cattle. For example proportion of hind-limb muscle in horse is steady to 38% whereas it falls from 38 to 33% in Limousin breed at a comparable stage of development (Robelin *et al.,* 1977).

Comparing carcass composition in adult to 12 months colt proportion of muscle, bone and adipose tissue are –1.0, +1.3 and +1.0% respectively. But it is highly depending on the finishing period before slaughter.

Effect of sex

Empty live body weight of colts is on average 70 kg higher than that of fillies. As a result carcass weight is 50 kg higher. But true yield is very close: 70% for both sexes (Table 4).

Table 4. Effect of sex and French draft breeds or body weight, carcass characteristics of horses slaughtered between 12 to 30 months (Robelin et al., 1984).

		Live				Internal thoracic adipose tissue[4] (kg)	Tissue composition of the carcass			Muscle/bone
		Body weight (kg)	Empty body weight[1] (kg)	Carcass[2] weight (kg)	True yield[3] (%)		Muscle (%)	Adipose tissues[5] (%)	Bone (%)	
Age (months)	12	483.2	439.6	313.4	71.2	3.86	70.1	10.9	15.6	4.48
	18	572.7	474.0	328.9	69.3	2.97	71.8	9.4	16.1	4.46
	24	626.8	539.6	382.7	70.9	5.94	69.8	12.9	14.9	4.69
	30	735.3	622.0	440.8	70.9	9.86	69.0	14.2	14.5	4.81
Sex[6]	Male	628.6	535.5	377.2	70.4	5.46	70.7	11.0	15.6	4.56
	Female	558.3	485.0	343.6	70.8	5.28	69.7	12.5	15.1	4.63
Breeds[7]	Ardennaise	599.8	516.9	362.6	70.1	6.97	69.6	12.9	14.9	4.69
	Boulonnaise	583.1	498.5	352.3	70.6	3.99	71.6	9.2	16.4	4.38
	Bretonne	568.9	480.9	338.9	70.5	4.19	70.9	10.9	15.5	4.57
	Comtoise	570.3	492.3	347.0	70.4	7.15	68.5	14.3	14.3	4.80
	Percheronne	658.9	572.0	407.1	71.7	4.37	70.9	10.8	15.7	4.52

[1] Empty body weight = live body weight – digestive content weight.
[2] Hot carcass weight.
[3] Hot carcass weight/empty body weight.
[4] Highly relevant criteria for predicting body condition score (Martin-Rosset *et al*.,1983).
[5] Adipose tissues: fat.
[6] All ages together.
[7] French draft breeds.

Table 5. Variations of the different adipose tissues proportions in the horse body of French draft breeds at 6 and 30 months of age (Robelin et al., 1984).

	Age at slaughter	
	6 months	30 months
Total adipose tissues[1] (% empty body weight)	7.6	11.6
Subcutaneous adipose tissues (% total adipose tissues)	23.1	30.8
Intermuscular adipose tissues (% total adipose tissues)	51.4	43.8
Internal thoracic adipose tissues (% total adipose tissues)	10.9	13.1
All abdominal adipose tissues (% total adipose tissues)	11.9	14.3

[1] Adipose tissues: fat.

Proportion of adipose tissue in the carcass is lower in colts than in fillies (11.0 vs 12.5%) and the proportion of muscle is higher in colt (70.7 vs 69.7%) but the differences are not significant (Table 4). The relative weights of muscle areas are very similar in both sexes (Table 6).

Muscles of colts are harder than those of fillies (Table 7). But as the content and the solubility of collagen are very close the discrepancy might be explained only by a difference in structural organization of the meat. It is coarser in colt than in fillies.

Effect of breeds

High size breed such as *Percherons* provides heavier carcass (407 kg) than the other breeds (339 to 363 kg) and true yield is slightly higher 71 vs 70% but not statistically significant (Table 4).

Proportion of adipose tissues and muscles in the carcass of *Comtois* and *Ardennais* breeds are 36% (relative value) higher and 31% (relative value) lower than those of other breeds (Table 4) respectively. Genetic differences are much higher in cattle: 50% (relative value) in young bulls (Geay and Malterre, 1973). As a result the same fat content in the carcass is reached at significantly different body weight for the breeds: 471 kg for *Comtois*; 486 kg for *Bretons*; 508 kg for *Ardennais*; 519 kg for *Boulonnais* and 580 kg for *Percherons* (Martin-Rosset et al., 1980). These discrepancies are much lower in horse than in beef cattle.

Physicochemical characteristic of muscles are slightly different between breeds namely toughness but no discrepancy is statistically significant (Table 7). Genetic discrepancies of collagen content and colour are much higher in young bulls (Boccard et al., 1979).

Effect of diet

Comparing colts fattened between 6 to 12 months with maize silage or good hay fed free choice and supplemented with the same appropriate amount of concentrate to meet the requirement of the same daily gain: carcass weight, true yield are slightly higher with maize silage based diet than with hay based diet; and fat and muscle contents are higher (+ 2 points) and lower (- 2.5 points) in colt

Table 6. Weight proportions of the main muscular areas in horse empty body weight: effects of sex and French draft breeds (Robelin et al., 1984).

		Muscles of the back[1] (%)	Muscles of the forelimb[1] (%)	Muscles of the hindlimb[1] (%)
Age (months)	12	12.0	13.5	38.0
	18	12.3	13.7	38.3
	24	12.0	13.7	38.2
	30	11.9	13.3	37.8
Sex[2]	Male	12.1	13.6	37.8
	Female	12.0	13.5	38.4
Breeds[2]	Ardennais	12.1	13.7	38.3
	Boulonnais	12.2	13.5	38.3
	Breton	11.9	13.5	38.2
	Comtois	12.0	13.5	37.6
	Percheron	12.2	13.5	38.0

[1] In % of total muscles.
[2] All ages together.

Table 7. Effects of breed and sex on muscles characteristics in horses of French draft breeds (fitted mean provided by variance analysis of all the data obtained at 12, 18, 24 and 30 months of age): shear force, collagen content and solubility and colour (Robelin et al., 1984).

		Longissimus dorsi	Profundus pectoralis	Triceps brachii caput laterale	Semi tendinosus	Transversus abdominis
Heavy breeds						
Ardennais	A	1.58	4.85	3.96	4.18	1.96
	B	2.22	5.21	5.08	4.07	2.61
	C	0.357	0.317	-	0.385	-
	D	12.82	20.78	16.38	9.17	13.69
Boulonnais	A	1.53	4.70	3.55	3.76	1.70
	B	2.60	5.12	4.90	4.07	2.71
	C	0.372	0.315	-	0.398	-
	D	13.43	17.87	15.78	9.28	12.89
Breton	A	1.33	4.53	3.95	3.92	1.78
	B	2.58	5.41	5.08	4.10	2.58
	C	0.360	0.330	-	0.379	-
	D	13.98	18.63	16.46	10.02	14.03
Comtois	A	1.29	4.48	3.21	3.55	1.45
	B	2.22	4.72	4.56	3.53	2.41
	C	0.339	0.291	-	0.365	-
	D	12.67	17.41	16.05	9.31	12.38
Percheron	A	1.18	4.79	3.20	3.38	1.55
	B	2.87	5.61	4.68	3.78	2.71
	C	0.339	0.306	-	0.365	-
	D	12.36	17.76	15.08	9.23	12.43
Sex						
Males	A	1.54	4.91	3.66	3.95	1.76
	B	2.48	5.27	5.04	3.90	2.65
	C	0.350	0.305	-	0.370	-
	D	13.95	20.20	17.01	10.17	13.32
Females	A	1.18	4.38	3.44	3.50	1.59
	B	2.51	5.12	4.63	3.89	2.54
	C	0.356	0.318	-	0.388	-
	D	11.98	16.40	14.71	8.50	12.73

[1] A = shear force (daN); B = collagen (N of hydroxyprolin / total N: g/kg); C = solubility of collagen (kg/kg); D = colour (heminic iron: γ/g muscles).

fed maize silage based diet than in colt fed hay diet; but the difference are not statistically different (Table 8). Only the solubility of collagen is higher with maize silage based diet (Table 8).

Light breeds: experimental data

The characteristics of carcasses were described in gelding between 2 to 14 years of age in light French breeds culled then slaughtered in experiments carried out in France at INRA (Table 9). The horses were fed to reach different fat content evaluated by a body score method (INRA *et al.*, 1997). Then tissue composition was determined using total anatomical dissection of the carcass according to the same method described in draught breeds by Martin-Rosset *et al.* (1983).

Table 8. Carcass and muscles characteristics in one year old and in adult horses of French draft breeds. Effect of diet (Robelin et al., 1984).[1]

	One year horse		Adult horse (broken mare)
Diets	Hay	Maize silage	Hay
	Free choice + 4 kg concentrate	Free choice + 4 kg concentrate	Free choice + 2 kg concentrate
Number	10	10	8
Body weight (kg)	486.4	480.1	709.0
Hot carcass weight (kg)	320.2	306.6	420.3
True yield (%)	71.7	70.8	68.3
Tissue composition (%)			
Muscles	71.7	69.1	69.4
Bone	15.9	15.8	17.2
Fat	9.8	12.0	11.8
Tougthness (kg/cm^2)			
Longissimus dorsi	1.48	1.28	-
Profundus pectoralis	4.10	4.21	-
Collagen content: N of hydroxyprolin/total N (g/kg)			
Longissimus dorsi	2.16	2.12	1.95[2]
Profundus pectoralis	4.62	4.34	7.88[2]
Solubility of collagen (%)			
Longissimus dorsi	0.43a	0.52b	-
Profundus pectoralis	0.37a	0.44b	-
Colour: iron (µg/g fresh)			
Longissimus dorsi	11.61	12.26	16.20[2]
Profundus pectoralis	16.30	15.49	21.30[2]

[1] The means on the same line with different letters (a, b, c) are statistically different (*P*£0.05).
[2] Average data for 4 animals.

Caracass weight is of course lower in light breeds (500 kg average live body weight) than in young draft breeds (605 kg average livebody weight: Table 4 and 9). But difference in true yield of carcass is weak 69.7 and 70.6% in light and draft breeds respectively. Muscles percentage is on average 3 points higher in light breeds than in draft breeds. This trend is confirmed when comparing adult light breeds (Table 9) with adult draft breed (Table 8). However, this discrepancy should be considered very carefully for fat content of carcass of breeds was quite different: 7.4% (Table 9) and 11.8% (Table 8) or 11.9% (Table 4) respectively. Experimental goals were different. All young draft breed animals were fed to meet optimal carcass fat content (see Martin-Rosset *et al.*, 1985b) whereas all adult light breed animals were fed to reach different body fat content o design a body condition scoring method (see Martin-Rosset *et al.*, 2008). In addition, draft breed animals are 12 to 30 months old while light breed animals are 2 to 14 years old. Bones percentages are quite close (16.2 vs 15.3%) and the ratios muscles/bones are also very similar (4.55 vs 4.61). Comparison between adult light breeds (Table 9) and adult draft breeds (Table 8) can be attempted. The differences are oriented in the same way as for the previous comparison. Unfortunately, the breeds are not exactly compared at the same fat content.

Table 9. Live body weight and carcass characteristics of light breed horses culled and slaughtered between 2 and 14 years old (n=20) (adapted from Martin-Rosset et al., 2008).[1,2]

	Live body weight (kg)	Empty body weight[3] (kg)	Hot carcass weight[4] (kg)	True yield[5] (kg)	Carcass composition						Muscle/bone ratio
					fat		muscles		bone		
					kg	%	kg	%	kg	%	
Mean	500.7	440.1	307.0	69.7	34.3	7.4	223.6	73.1	46.7	16.2	4.55
SD	60.0	55.2	43.8	1.5	19.7	3.6	29.6	2.9	5.4	1.5	0.39
CV (%)[6]	12.0	12.5	14.3	2.2	57.4	48.6	13.2	4.0	11.6	9.3	8.6

[1] Horse breeds: French saddle – French trotters.

[2] Horses were prepared to be slaughtered at very different body condition score (1.0 to 4.5 according, INRA scale, 0-5) to study and to establish relationship between body condition and body composition determined using total anatomical dissection (Martin-Rosset et al., 2008).

[3] Empty body weight = live body weight – digesta content weight.

[4] Hot carcass weight at slaughter.

[5] True yield = carcass weight / empty body weight.

[6] CV: coefficient of variation.

As a result, the data obtained in light breeds are very informative. Unfortunately, comparison between light and draft breeds is not yet conclusive for these breeds should be compared within the same range of age and even better at the same fat content (or body condition score). It has been already showed in draft breeds that breeds should be compared accordingly (Table 4) to highlight differences.

Methods for predicting body condition score and tissues carcass composition

Body condition score (BCS) of horse before slaughtering can be assessed with a standardized manual method performed in France (INRA et al., 1997; Martin-Rosset, 2015). This BCS method has been validated by tissue composition measured after total anatomical dissection (Martin-Rosset et al., 2008).

Tissues composition of the carcass can be assessed from the total anatomical dissection of either the shoulder or the 14th rib. (Martin-Rosset et al., 1985a). But in routine slaughter house, the quality of carcass can be assessed for young and adult horse through an official classification table stated by OFIVAL, 1985 updated by INTERBEV, 2015, www.interbev.fr, the official interprofessional organization in France using the conformation and the importance and the repartition of subcutaneous and internal thoracic adipose tissues (Roy and Dumont, 1976). Optimal fat content of the carcass can be easily evaluated after slaughtering using weight of internal thoracic fat for each class of age (Table 2).

Feeding: requirements and recommended allowances

Requirements and daily feed allowances of all these breeds have been determined for different body weight, age, sex and daily gain to meet expected performance by INRA (Jarrige and Martin-Rosset,

1984) then updated in 1990 (Martin-Rosset, 1990) and finally in 2012. The daily nutrient and intake recommended are either displayed in tables to be used directly by end-users or they can be evaluated using models where expected performance are related to nutrient intake (Martin-Rosset, 2015). Recommendations are also proposed for light breeds.

Meat characteristics

Evolution of horse meat after slaughter

As in other animal species 3 steps take place during the maturation process for muscle to be transformed in meat: quivering, rigid, mature stages. Rigor mortis and maturation occur between the 2 first and the 2 last steps respectively.

Maximum cadaveric rigidity occurs in horse and beef meat respectively 48 h and 24 h after slaughter (Ulyanof and Tuleuov, 1976) as glycogen content of horse muscles is much higher than in beef muscles. But maturation of horse meat is faster (4 to 5 days) than that of beef meat (5 à 8 days) as a result horse meat can be consumed in a fresh stage (Boccard, 1975; Pavloskij and Palov, 1979). Horse meat must be sold quickly after being preserved at low temperature in respect of the risk of microbial contamination due to high glycogen and non protein nitrogen content (Cantoni *et al.*, 1979; Cattaneo et *al.*, 1979).

Physico chemical characteristics of horse meat and organoleptic qualities

Draft breeds

The tenderness and the colour were measured on 3 hard muscles *(Semi-tendinosus, Pectoralis profundus, Triceps brachii caput laterale)* and 2 tender muscles (*Transversus abdominis, Longissimus dorsii*) (Figure 3: Boccard, 1975; Robelin *et al.*, 1984).

The toughness of tender muscles is steady whatever it may rise for hard muscles between 12 to 30 months (Figure 3A). The collagen content is twice time higher in the hard muscle: *Pectoralis profundus* than in the tender muscle, *Longissimus dorsi*. For the 5 muscles collagen contents rise very much between 12 months to 18 months then they keep on being steady (Figure 3B). The initial rise takes place in the same time of puberty in males as in cattle (Boccard *et al.*, 1979). But the collagen content in horse muscle is on average higher than in cattle for example: N hydroxipoline / N total are 2.1. vs 1.8% in *Longissimus dorsi* in horse and in cattle respectively (Boccard *et al.*, 1979; Robelin *et al.*, 1984). The solubility of the collagen decreases with the age from 40-50 to 22-27% between 12 to 30 months (Figure 3C). As a result the tenderness of the meat is reduced.

Comparing adult to young horses, the collagen content of muscles is slightly lower for the *Longissimus dorsi* but it is very much higher for the *Pectoralis profundus* in adult than in young horses (Table 7 and 8). As a result the higher expected tenderness of meat provided with old horse seems to be not confirmed in adult draft breeds used only for breeding.

The colour is due to the myoglobine content of muscles which is measured as heminic iron content. The colour rises with the age (Figure 3D), which is quite similar to what occur in cattle (Renerre, 1982). In adult horse the colour is higher than in young horse (Table 7 and 8).

The flavour is very soft due to high glycogen content: 0.5 to 3.0 times higher than in beef meat (Pitre, 1975). Some unsaturated fatty acids would be suspected to contribute to the flavour as well.

Light breeds

Influence of age on chemical, physical characteristics on preservation and sensory properties of horse meat have been very well highlighted in Switzerland at the Research Federal center (Agroscope, Posieux, Vaud) by an experiment using Franches-Montagnes horse breed, a multipurpose breed, 550 kg adult body weight (Table 10). Meat of 30 months-old-animals was compared to that of 7 months-old (slaughtered at weaning),and to that of more 5 years-old-culled horses. It has been concluded that the meat of 30 months-old-animals is the good optimum in respect of iron content, red colour and redness, collagen content and solubility and related shear force (Table 10). Meat is easier to preserve and palatibilities properties are optimal, namely tenderness (Table 10).

Characteristics of horse meat for human consumption

Consumer's preferences

Consumer 's preferences have been tested in the scope of the experiment carried out with Franches-Montagnes breed using a panel approach which included both scoring test using qualitative criteria and non-structured scale and preference tests (Dufey, 2001). Comparing horse meat from 30 months-old animals with weaned foals 7 months-old and adult culled over 5 years-old, it is concluded that preference rate is 50% for meat of 30 months-old animals whereas the rate is 40% for other types of horses (Figure 4). It is consistent with chemical, physical and sensory properties measured in the same experiment (Table 10).

Nutritional characteristics

The chemical composition of horse meat is different to that of other herbivores and pig (Table 11). Water, protein, glycogen, iron and hydrosoluble vitamins are higher whereas lipids, sodium and liposoluble vitamins are lower in horse meat than in ruminants and pig meat. As a result nutritional characteristics of horse meat are specific.

Table 10. Influence of age on chemical, physical and subsequent preservation and sensory properties in muscle longissimus thoracis (Dufey, 2001).[1]

Characteristics	Unit	Age		
		7 months	30 months	Adult
Heminic iron	mg/100 g	1.09[c]	1.94[a]	1.61[b]
Redness	Index	13.70[a]	13.72[a]	12.61[b]
Collagen	mg/100 g	392	355	375
Collagen solubility	%	46.1[a]	27.2[b]	15.3[c]
Shear force	kg	2.77[b]	3.22[a,b]	3.37[a]
Preservation				
Aging	%	6.29[a]	4.51[b]	4.03[b]
Freezing	%	7.33[a]	5.91[b]	6.06[b]
Palatability traits				
Flavour	pts	4.61[a]	4.10[b]	4.11[a]
Juiciness	pts	3.49[b]	4.09[a]	4.04[a]
Tenderness	pts	4.89[a]	4.54[a]	3.68[b]

[1] The means on the same line with different superscripts (a, b, c) are statistically different ($P≤0.05$).

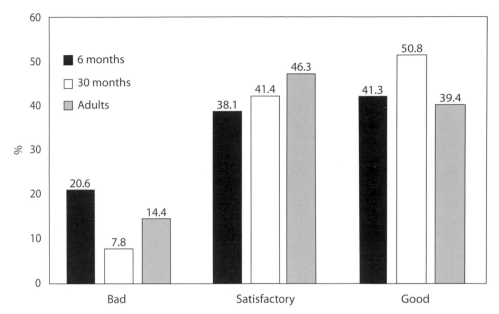

Figure 4. Panel's preference of consumers (%) for meat of 7 months, 30 months old and adult horses, evaluated according three classes: bad; satisfactory and good (Dufey, 2001).

Table 11. Chemical composition of meat in different animal farm species for 100 g of edible meat (Pitre, 1975).

Components	Cattle	Calf	Pig	Sheep	Horse
Water (g)	53.0-71.6	68.0-70.0	52.0-60.0	58.0-64.0	73.2-75.2
Proteins (g)	15.5-19.3	19.1-19.5	14.6-16.6	15.6-18.0	21.6-23.3
Lipids (g)	13.0-28.0	9.0-12.0	23.0-32.0	17.0-26.0	0.5-3.0
Glycogen and derives (g)	0.7-2.0	–	0.2-0.8	–	2.4

Calorific value is low (140 kcal or 580 kJ/100 g of raw meat in adult horse (Badiani *et al.,* 1997) due to low lipid content of carcass, high relative proportion of subcutaneous and internal adipose tissue to total adipose tissues of the carcass which are removed during the preparation of horse meat for selling by the butcher, and finally due to lipids composition. Triglycérides provide the most important fraction of the lipids (Orlov *et al.,* 1985). Comparing horse meat to beef meat and to a lesser extend to pig meat, unsaturated fatty acids are much higher (Pitre, 1975). According to Baudon (1985) and Catalano and Quarantelli (1979), unsaturated fatty acids account for 60 to 70% of total lipids in horse meat. Melting point of horse meat lipids is low in respect of such a composition which is a medical advantage for preventing cardiovascular diseases (Baudon, 1985). Glycogen content is higher in horse than in cattle or pig (Table 11). But glycogen content is higher in adult horse than in young horse (Dufey, 2001).

Cholesterol content averages 45.7 mg/100 g fresh (longissimus thoracis) and it is very steady with age (Dufey, 2001).In adult horses cholesterol content ranges between 49-73 mg/100 g fresh on the basis of composite samples representative of carcass (Badiani *et al.*, 1997).

Essential amino acids content of horse meat is higher namely in: lysine, leucine and to a lesser extend in arginine, acids glutamique and aspartique, histidine than in beef, ovine meat (Vervack *et al.,* 1977).

Sodium, potassium and iron content average 74, 331 and 3.9 mg/100 g fresh raw horsemeat respectively (Badiani *et al.,* 1997). As sodium and iron content of horse meat are 3 to 4 times lower and 2 to 3 times higher respectively, than in other beef, ovine, pig meat. Horse meat can be designed for human nutrition recommending hyposdic diet or for reducing anaemia. Indeed, 100 g of horse meat on raw basis provide 14.8 and 27.8% of the estimated safe and adequate dietary intake for sodium and iron respectively on the basis of the RDAs suggested by the European Union Council (EEC, 1990), (Badiani *et al.,* 1997).

Hydrosoluble vitamins content is as far high as the lipids content is low. C and PP vitamins contents are satisfactory (Khamidullina, 1984; Baudon, 1985).

Chemical composition of horse meat has been recently detailed in different typical pieces of meat in the scope of a large study managed by the French Interprofessional Meat Center (CIV in French: Centre interprofessionnel des viandes, Paris; www.civ-viande.org). This study was planned for that purpose and analysis were carried out at INRA (Table 12 13 and14).

Dry matter content is 9-10% higher in sirloin (SI) and rib eye steak (RES) than in topside (TO) mainly in adult horse (Table 12). Total lipids or total fatty acids content are on average 2.5 and 1.6 times respectively higher in SI and RES than in TO in adult and young horse. Total iron content is between 1.5 to 1.9 times higher in adult than in young horse according to the pieces of meat. These observations are confirmed by heminic iron content. Zinc content is only higher in adult's TO piece than in young horse. Selenium is on average 35% higher in adult horse than in young horse but not different between pieces of meat in each category of age. Vitamin B3 content is very high in adult and young horse whatever the pieces of meat. There is no influence of age on and piece of meat on vitamin B6 content whereas vitamin B12 content is more than twice time higher in adult than in young horse in all pieces of meat. Chemical compositions rely mainly on age of horse and to a lesser extend to piece of meat according the components. But interaction between should not be excluded.

In the same study fatty acids composition was detailed according to age and piece of meat and expressed in percentage of total fatty acids content (Table 13). The total of saturated fatty acids (SFA) falls within 38-40% whatever age and piece of meat. And there is no difference in linear or branched saturated fatty acids percentage according to the same criteria. The total of mean unsaturated fatty acids (MUFA) percentage is either very close to total of SFA in adult whatever piece of meat. In contrast, the total MUFA percentage is lower than total SFA in young horse whatever piece of meat. The total MUFA *cis* percentage is much higher than MUFA *trans* percentage in adult and in young horse for all pieces of meat. MUFA *cis* percentage is 1.4 time lower in young horse than in adult horse whatever the pieces of meat while MUFA *trans* percentage is slightly lower (18-20 points) for all pieces of meat in young horse than in adult. Again fatty acids composition relies mainly on age and to a lesser extent on piece of meat. But interaction between should not be excluded.

Difference of essential amino acids content is mostly lower than 10% according to the pieces of meat (Table 14). But thrytophane content is 10-12% lower in TO and RES than in SI. And methionine or phenylalanine is 26 and 10% lower in SI and RES respectively. Amino acids composition seems to be less influenced than fatty acids composition.

Hygienic characteristics

Horses are good indicators for cadmium and lead pollution as these heavy metals are accumulated in liver and kidney (Hecht, 1984; Holm, 1979; Salmi and Hirn, 1981). As a result, cadmium and

Table 12. Dry matter, crude protein, lipids, total fatty acids, mineral (iron: total and heminic, zinc, selenium) and vitamins B (B3, B6, B12) content of three pieces of meat in horse: culled saddle adult (A, n=8) and in young draft horse (YH, n=8) (Bauchart et al., 2008).[1]

		Topside	Sirloin	Rib eye steak
Main nutrients	Dry matter (g/100 g fresh)			
	A	25.87	28.47	28.23
	YH	25.25	26.08	25.67
	Total protein (g/100 g fresh)			
	A	23.00	21.95	22.39
	YH	22.39	22.46	21.79
	Total lipids (g/100 g fresh)			
	A	2.13	5.17	4.95
	YH	1.24	1.98	1.73
	Total fatty acids (g/100 g fresh)			
	A	1.64	4.39	4.15
	YH	0.89	1.58	1.35
Main minerals	Total iron (mg/100 g fresh)			
	A	3.99	3.30	3.38
	YH	2.13	1.80	2.30
	Heminic iron (µg/g fresh)			
	A	2.31	1.89	2.52
	YH	1.27	1.09	1.10
	Zinc (mg/100 g fresh)			
	A	2.73	1.86	2.52
	YH	1.90	1.97	2.50
	Selenium (µg/100 g fresh)			
	A	6.16	6.21	6.29
	YH	4.49	4.49	4.45
Vitamins – B group	Vit B3 (mg/100 g fresh)			
	A	5.28	5.22	5.55
	YH	5.50	5.98	6.75
	Vit B6 (mg/100 g fresh)			
	A	0.69	0.69	0.66
	YH	0.71	0.63	0.66
	Vit B12 (µg/100 g fresh)			
	A	2.20	2.11	2.04
	YH	1.06	0.91	0.84

[1] Topside: Musculus gracilis; sirloin: longissimus dorsi; rib eye steak: longissimus dorsi; Spinalis thoracis; Multifidus lumborum et thoracis; Intercostalis.

lead concentration in these organs used to be higher to the admitted standards. Conversely, there are no risks concerning other heavy metals and trace elements as their contents are lower than in other meat. In the mean time standards for horse meat in France are the same as for beef meat. The questions might be: do they have to be more specific, namely for cadmium and lead (Rossier and Berger, 1988)?

Table 13. Composition of total fatty acids (%) in saturated fatty acids (SFA) and monounsaturated (MUFA) of total lipids in three pieces of horse meat (topside, sirloin, rib-eye steak) (Bauchart et al., 2008).[1]

Fatty acids	Adult saddle breed horse (n=8)			Young draft breed horse (n=8)		
	Topside	Sirloin	Rib-eye steak	Topside	Sirloin	Rib-eye steak
Saturated linear	38.7	39.5	39.6	37.4	39.0	37.5
Saturated ramified	0.11	0.14	0.13	0.10	0.10	0.09
SFA total	38.8	39.6	39.7	37.5	39.1	37.6
MUFA cis	33.6	41.4	39.4	23.0	28.7	28.0
MUFA trans	0.58	0.69	0.65	0.43	0.55	0.53
MUFA total	34.1	42.0	40.0	23.4	29.3	28.6

[1] Topside: Musculus gracilis; sirloin: longissimus dorsi; rib eye steak: longissimus dorsi; Spinalis thoracis; Multifidus lumborum et thoracis; Intercostalis.

Table 14. Average content (g/100 g of fresh tissue) of essential amino acids (Isoleucine, Leucine, Lysine, Methionine, Phenylalanine, Threonine, Tryptophane, Valine) or semi essentials (Histidine) of three pieces of meat in draft breed horse (topside, sirloin, rib eye steak) in horse (Bauchart et al., 2008).[1]

	Topside	Sirloin	Rib-eye steak
Relevant amino acids			
Histidine	0.954	0.935	0.864
Isoleucine	1.074	1.127	1.053
Leucine	1.943	2.002	1.901
Lysine	1.807	1.805	1.702
Methionine	0.809	0.751	0.644
Phenylalanine	0.909	0.825	0.857
Threonine	1.077	1.073	1.018
Tryptophane	0.242	0.272	0.248
Valine	1.078	1.080	1.031

[1] Topside: Musculus gracilis; sirloin: longissimus dorsi; rib eye steak: longissimus dorsi; Spinalis thoracis; Multifidus lumborum et thoracis; Intercostalis.

Residue of pharmaceutical and industrial products such as antibiotics and pesticides respectively was not a critical point years ago (Madarena *et al.,* 1980; Meyrial, 1974). Is the situation today still the same? Certainly not It is increasingly clarified by new regulation.

Horse meat is very scarcely subjected to microbial and parasite contaminations (Catsaras *et al.,* 1974a,b; Corsalini, 1976; Magras *et al.,* 1997; Rossier and Berger, 1988). But as horse meat might be consumed either raw chopped steak or slightly grilled, hygiene conditions for meat preparation should be very strict as reviewed by Rossier and Berger, 1988. Outbreaks of Trichinosis were pointed out in France and in Italy in the years 1975-1985 (Bouree *et al.,* 1977, 1979; Bussieras, 1976; Mantovani *et al.,* 1976, 1979). But Trichinosis should be considered as fairly exceptional (Busserias, 1976) as it is transmitted only through carnivorism (Bouree *et al.*, 1976). Nowadays, importations of horse meat into France are subjected to previous detection of Trichina according to EU guidelines since October 25th 1985. Whenever, live horse importations into France are subjected to this control in the slaughter house (Champalle, 1987).

Technological horse meat transformation

Horse meat requires a specific preparation from the butcher. All muscles should be dissected in different piece very well defined (see www.civ-viande.org). Then horse meat can be consumed raw, cooked and included in delicatessen. Organoleptics characteristics of horse meat would be well preserved after electrostimulation and lyophilisation (Calcano *et al.*, 1978; Dolatowski, 1985a). In Central European and Eastern Countries different technological process are implemented to improve the preparation of horse meat delicatessen: grinding (Tuleuov and Billalova, 1972; Tuleuov *et al.*, 1974) presalted (Cantoni *et al.*, 1979b; Pavloski *et al.*, 1972) adjuved with blood plasma (Dolatowski, 1985b; Tuleuov *et al.*, 1976). Horse meat can be identified in foods products (Babiker *et al.*, 1981; Casa *et al.*, 1985).

Discussion and conclusions

In France, a system of extensive production of herds of draft breeds mares can be efficient if the main technical factors concerning profitability are mastered: numerical productivity – colt weight at weaning and the cost of feeding the mare. However, an extensive type of husbandry has a limited productivity per animal and per ha. Hence it is mainly developed in France in extensive zones with low agricultural populations and where large areas are available.

Before 1970 most draft mares were traditionally raised in lowland grassland or mixed farming areas. The maintenance of draft mares in these regions can be thought, when grassland production is intensified. However, recent changes point out numerous technical and economic issues which have been little studied up to the present. The 'mare-foal couple is a modest transformer of feeds compared with other more productive species with which they are in direct competition in these regions. However, the horse can, no doubt, provide complimentary meat production by the only fact that it grazes turf much shorter than ruminants and feeds on their refusals, and that it can make use of poor areas unusable for ruminants (Fleurance *et al.*, 2012; Osoro *et al.*, 2012). It could contribute to the diversification of animal production in farms as it has been evaluated (REFERences network, 2009-2013).

In the other countries such as Belgium heavy horses are not specialized for horse meat production even consumption is still significant. For most part supply demands are still met through importation. In Italy, there is a trend towards specialized production since the 1980's (Badiani *et al.*, 1993; Catalano *et al.*, 1986). Particularly the production of colts ready for fattening at weaning in marginal lands which are important in this country: 60% of agricultural and forestry surface (Catalano and Miraglia, 1985; Catalano *et al.*, 1986). But such a production is not enough to meet the needs. Italy import colts of heavy breeds from France and light breeds from other countries. Spain is a category of its own. The equine population is very high: 600,000. But mules and asses represent more than 50%. There are few draft horses which are managed in the marginal lands in Pyrenean Mountains (Osoro *et al.*, 1012) to provide horse meat to Barcelona's market.

In all these countries, draft breeds are involved in grassland and heathland ecosystems exploitation and preservation where they are favourably impacting socio economy of these rural areas (Miraglia *et al.*, 2006; Vial *et al.*, 2012). In addition, the release of enteric methane by equids is far less than in ruminants (Vermorel *et al.*, 2008). It has been calculated as 20,202 tonnes from a total of 975,000 animals in France. This gives an emission number of 20.7 kg/animal/year. For the lactating mare the number is higher: 29.7 kg, which represents 34% of the value calculated for the milking cow Overall, the total enteric methane emissions from equids in relation to all farm animals accounts for 1.5% compared to 90% for ruminants in France (Martin-Rosset *et al.,* 2012).

In the countries managing light horses for leisure, sports and racing: all the EU countries are concerned but mainly France, Germany, United Kingdom, Eire and the Netherlands but increasingly Spain as well.

These countries have high number of light horses which have increased over the last 15 years as the number of leisure riders and sportsmen have increased so much. In addition horse racing has been further developed recently as well.

These horses are generally managed during the breeding period in association with ruminants, according to husbandry and feeding techniques that have been tested in experiments in France (Micol and Martin-Rosset, 1995; Morhain, 2011; REFERences network, 2009-2013; Trillaud-Geyl and Martin-Rosset, 2011) and observed to some extend in the other European countries.

The culling of these horses at the end of the working life is a great source of supply. It accounts for 70% of the internal consumption in France because this horse meat, locally supplied but also largely imported, is less expensive. And it is nearly 100% in non consuming countries where it is exported. The proportion of light horse breeds in horse meat consumption must increase with respect to the relative negative evolution of draft horse breeds number to light horse breeds and for economical reasons given: the rising pressure of selection and to the age at the first performance test (for example two years against three years for trotters) and the number of performances test. Integration into EU of new countries: Poland where there is still large population of horses and Hungarian or Romania with a significant population increases the potential supply for horse meat.

Knowledge about carcass and meat characteristics needs to be refined. And significant differences, if any, comparing with draft breeds, are not yet definitely established.

Horse meat consumption represents only a small part of total meat consumption for example, in France is limited. Consumption is either of red meat from adult horses (mainly of light breeds) or old colts or, paler meat from colts raised for the purpose, of draft breeds and slaughtered between 6 and 24 months of age.

The reasons for limited consumption seem to be at least for France:
- The symbolism attached the horses (noble animal, friend of man).
- The individual taste: horse meat is considered as lacting flavour and requiring considerable seasoning. It is also judged to have poorer conservation quality than other meats. Its high soluble nitrogen and glycogen contents could result in an increased risk of contamination. The high iron content causes a fast change in coloration. The chain from slaughter to consumption must therefore be well controlled.
- Old and out of date laws forbidding consumption in cantines, restaurants, etc.
- Finally, commercialisation difficulties related to sales organisation: generally by specialized butchers, whereas super and hypermarkets sell very little horse meat, and to high retail prices, above those of beef in recent years which are not justified by preparation costs.

In contrast, horse meat has important dietetic qualities: tenderness, low fat content, high protein content, and technology for preservation has been improved those least decades (Rossier and Berger, 1988). And they are now rather well established and promoted by the Interprofessional Meat Centre, so-called CIV in France (www.civ-viande.org).

Horse meat production from draft breeds cannot be an alternative production of other farm animal production; however, it could be complementary to beef or milk production in highland areas and in extensive grassland zones as well as pointed out by Martin-Rosset (1987) in a report provided for the European Commission. The difficulties to be resolved pertain to management both to ensure complementary production and good technicity. And the extension or maintenance of draft mare herd has been steadily fixed in relation to the market and to saddle horse production which is increasing in most of industrialized countries.

References

Agabriel, J., C. Trillaud-Geyl, W. Martin-Rosset and M. Jussiaux, 1982. Utilisation de l'ensilage de mals par le poulain de boucherie. INRA Productions Animales 49: 5-13.

Babiker, S., P.A. Glover and R.A. Lawrie, 1981. Improved methodology for the electrophoretic determination of horse meat in heated foodstuffs. Meat Science 5: 473-477.

Badiani, A., M. Manfredini and N. Nanni, 1993. Qualita della carcassa e della carne di puledri lattoni. Zootecnica Nutrizione. Animate 19: 23-31.

Badiani, A., N. Nanni, P.P. Gatta, B. Tolomelli and M. Manfredini, 1997. Nutrient profil of horsemeat. Food Composition and Analysis 10: 254-269.

Bauchart, D., F. Chantelot, A. Thomas and L. Wimel, 2008. Caractéristiques nutritionnelles des viandes de cheval de réforme et de poulain de trait. Centre d'Information des Viandes, Paris. France.

Baudon, D., 1985. Viande de cheval. Le concours medical. 107-200.

Bigot, G., A. Celie, S. Deminguet, E. Perrte, J. Pavie and N. Turpin, 2011. Managing grassland in sport horse breeding farms in lower Normandy. Fourrages 207: 231-240.

Boccard, A., 1975. La viande de cheval. INRA Productions Animales 21: 53-57.

Boccard, R., R.T. Naude, D.E. Cronje, M.C. Smit, H.J. Venter and E.J. Rossouw, 1979. The influence of age, sex and breed of cattle on their muscle characteristics. Meat Science 3: 261-280.

Bouree, P., G. Kouchner, A. Gascon, J. Fruchter, J. Passeron and J.B. Bouvier, 1977. Trichinose: bilan de l'épidémie de janvier 1976 dans la banlieue sud de Paris. Annales de Médicine Interne 128: 647-654.

Bouree, P., J.B. Bouvier, J. Passeron, P. Galanaud and J. Dormont, 1979. Outbreak of trichinosis near Paris. British Medical Journal 1: 1047-1049.

Bussieras, J., 1976. L'épidémiologie de la trichinose. Recueil de Médecine Vétérinaire 152: 229-234.

Calcano, L., C. Cantoni, M. Blanchi and G. Beretta, 1978. Lyophilisation des viandes de boeuf et de cheval. Industrie Alimentari 11: 827-830.

Cantoni, C., A. Aadaelli and G. Soncini, 1979. Pouvoir tampon du muscle de cheval. Archivio Veterinario Italiano 30: 44-45.

Cantoni, C., S. D'aubert, M. Perlasca, M. Blanchi and E. Proverbio, 1979. Production de saucisses 'wurstel' avec la viande de cheval. Archivio veterinario Italiano 30: 197-199.

Casa, C., J. Tormo, P.E. Hernandez and B. Sanz, 1985. The detection and partial characterization of of horse muscle proreins by immunoelectrophoresis in agarose gels. Meat Science 12: 31-37.

Catalano, A.L. and A. Quarantelli, 1979. Carcass characteristics and chemical composition of the meat from milk-fed foals. La Clinica Veterinaria 102: 498-506.

Catalano, A.L. and N. Miraglia, 1985. Exploitation of altitude's low productive pastures in wet areas. In proceedings of 36th Annual Meeting of EAAP. September 30 - October 3. Commission on Horse production. Kallithea, Greece.

Catalano, A.L., N. Miraglia, C. De Stefano and F. Martuzzi, 1986. Produzione di carne da cavalli di diverse categorie. Obiettivi e Documenti Veterinari 7: 69-73.

Catsaras, M., A.N. Gulistani and D.A.A. Mossel, 1974a. Contaminations superficielles des carcasses réfrigérées de bovins et de chevaux. Recueil de. Médecine Vétérinaire 150: 287-294.

Catsaras, M., R. Seynave and C. Sery, 1974b. Salmonella dans les boucheries, IV: contamination des boucheries de cheval. Bulletin de l'Académi Vétérinaire 47: 399-404.

Cattaneo, P., A. Aadaelli and C. Cantoni, 1979. Solubilité des fractions azotées du muscle de cheval. Archivio Veterinario Italiano 30: 47-48.

Champalle, E., 1987. Etude épidémiologique des cas de trichinose humaine observés en France en 1985. Mesures sanitaires vétérinaires. Cahiers de Nutrition et de Diététique 22: 479-482.

Corsalini, T., 1976. Rercherche sur la fréquence de présence des Salmonelles chez les chevaux importés et abattus régulièrement à Bari. Atti Della Societa Italiana delle Scienze Veterinarie 30: 551-553.

Dolatowski, Z.J., 1985a. Elektrostimulation und elektrisches feld. Einfluss auf veraenderungen im pferdefleisch nach dem schlachten. Fleischwirtschaft 65: 386-388.

Dolatowski, Z.J., 1985b. Influence on blood plasma on quality. Manufactured type products from horse meat. Fleischerei 36: 700-702.

Dufey, P.-A., 2001. Propiétés sensorielles et physico-chimiques de la viande de cheval issue de dfférentes catégories d'âge. In: 37ème Journée recherche Equine. Haras nationaux Editions, Paris, France, pp. 47-54.

ECUS, 2012. Tableau éconmique, statistique et graphique du cheval en France. Données 2011/2012.

Fleurance, G., N. Edouard, C. Collas, P. Duncan, A. Farruggia, R. Baumont, Ph. Lecomte and B. Dumont, 2012. How do horses graze pastures and affect the diversity of grassland ecosystems. In: Saastamoinen, M., M.J. Fradinho, S.A. Santos and N. Miraglia (eds.) Forages and grazing in horse nutrition. EAAP publication no. 132. Wageningen Academic Publishers, Wageningen, the Netherlands, pp. 147-162.

Geay, Y. and C. Malterre, 1973. Croissance, rendement et composition des carcasses de jeunes bovins de différentes races. Productions Animales 14: 17-20.

Hecht, H., 1984. Toxische und essentielle elemente in fleisch und organen von schlachtpferden. Fleischwirtschaft 64: 1113-1119.

Holm, J., 1979. Lead and cadmium levels in meat and organ samples from horses and possible effects on the health of the consumer's health. Fleischwirtschaft 59: 737-739.

INRA, HN and IE, 1997. Grille de notation de l'état corporel des chevaux de selle. Institut de l'Elevage Editions, Paris, France, 40 pp.

Jarrige, R. and W. Martin-Rosset (eds.), 1984. Le cheval: reproduction, sélection, alimentation, exploitation. INRA Editions, Versailles, France, 689 pp.

Khamidullina, L., 1984. Vitamins and trace elements in horse meat. Konevodstvo i Konnyi sport 12: 31-32.

Madarena, G., G. Dazzi and G. Campanini, 1980. Organochlorine pesticide residues in meat of various species. Meat Science 4: 157-166.

Magras, C., M. Fédérighi and C. Soulé, 1997. Les dangers pour la santé publique liés à la consommation de viande de cheval. Scientific and Technical Review of the Office International des Epizooties 16: 554-563.

Mantovani, A., I. Fllippini and A. Saccheti, 1976. Observations sur un foyer de trichinose humaine en Italie. Bulletin de l'Académi Vétérinaire 49: 213-217.

Mantovani, A., I. Fllippini and S. Bergomi, 1979. Indagini epidemiologiche su un focolaio di trichinellosi umana verificatosi in Italia. Parassitologia 21: 119-120.

Martin-Rosset, W. (ed.), 1990. Alimentation des chevaux. INRA Editions, Versailles, France, 232 pp.

Martin-Rosset, W. (ed.), 2015. Equine nutrition: INRA nutrient requirements, recommended allowances and feed tables. Wageningen Academic Publishers, Wageningen, the Netherlands, 624 pp.

Martin-Rosset, W. and C. Trillaud-Geyl, 2011. Pâturage associé des chevaux et des bovins sur des prairies permanentes: premiers résultats expérimentaux. Fourrages 207: 211-214.

Martin-Rosset, W. and M. Jussiaux, 1977. Production de poulains de boucherie. Productions Animales 29: 13-22.

Martin-Rosset, W., 1987. Horse for meat production. In: Agriculture: consequence of milk quotas and alternative animal enterprises. Commission of the European Communities, pp. 297-315.

Martin-Rosset, W., J. Vernet, H. Dubroeucq, A. Picard and M. Vermorel, 2008. Variation and prediction of fatness from body condition score in sport horses. In: Saastamoinen, M. and W. Martin-Rosset (eds.) Nutrition of the exercising horse. EAAP publication no. 125. Wageningen Academic Publishers, Wageningen, the Netherlands, pp. 167-176.

Martin-Rosset, W., M. Jussiaux, C. Trillaud-Geyl and J. Agabriel, 1985b. La production de viande chevaline en France. Systèmes d'élevage et de production. INRA Productions Animales 60: 31-41.

Martin-Rosset, W., M. Vermorel and G. Fleurance, 2012. Quantitative assessment of enteric methane emission and nitrogen excretion by equines. In: Saastamoinen, M., M.J. Fradinho, S.A. Santos and N. Miraglia (eds.) Forages and grazing in horse nutrition. EAAP Publication no. 132. Wageningen Academic Publishers, Wageningen, the Netherlands, pp. 485-492.

Martin-Rosset, W., R. Boccard, M. Jussiaux, J. Robelin and C. Trillaud-Geyl, 1980. Rendement et composition des carcasses de poulain de boucherie. Productions Animales 41: 57-64.

Martin-Rosset, W., R. Boccard, M. Jussiaux, J. Robelin and C. Trillaud-Geyl, 1983. Croissance relative des différents tissus, organes et régions corporelles entre 12 et 30 mois chez le cheval de boucherie de différentes races lourdes. Annales de Zootechnie 32: 153-174.

Martin-Rosset, W., R. Boccard, M. Jussiaux, J. Robelin and C. Trillaud-Geyl, 1985a. Estimation de la composition des carcasses de poulains de boucherie à partir de la composition de l'épaule ou d'un morceau monocostal prélevé au niveau de la 14e côte. Annales de Zootechnie 34: 77-84.

Meyrial, J., 1974. Identification et dosage du dichlorvos et de ses métabolites principaux chez le cheval. Thèse ENV, Lyon, France, 58 pp.

Micol, D. and W. Martin-Rosset, 1995. Feeding systems for horses on high forage diets in the temperate zones. In: Journet, M., E. Grenet, M.H. Farce, M. Theriez and C. Demarquilly (eds.) Recent developments in the nutrition of herbivores. Proceedings of 4th International Symposium Nutrition Herbivores Proceedings. INRA Editions, Versailles, France, pp. 569-584.

Miraglia, N., D. Burger, M. Kapron, J. Flanagan, B. Langlois and W. Martin-Rosset, 2006. Local animal resources and products in sustainable development: role and potential of equids. In: Rubino, R., L. Sepe, A. Dimitriadou and A. Gibon (eds.) Products quality based on local resources leading to improve sustainability. EAAP publication no. 118. Wageningen Academic Publishers, Wageningen, the Netherlands, pp. 217-233.

Morhain, B., 2011. Forage systems and feeding management in horses in different French regions. Fourrages 207: 155-164.

OFIVAL, 1985. Le marché des viandes et des produits avicoles. Tour du Maine Montparnasse. Paris, France, 184 pp.

Orlov, V., K. Servetnik, G.K. Chalaya and N.B. Zagibailova, 1985. Teneur et composition en acides gras des lipides des viandes de cheval et de chameau. Voprosy Pitanija 4: 71-76.

Osoro, K., L.M.M. Ferreira, U. Garcia, M. Rosa Garacia, A. Martinez and R. Celaya, 2012. Grazing systems and the role of horses in the heathland areas. In: Saastamoinen, M., M.J. Fradinho, S.A. Santos and N. Miraglia (eds.) Forages and grazing in horse nutrition. EAAP publication no. 132. Wageningen Academic Publishers, Wageningen, the Netherlands, pp. 137-146.

Pavlovski, P., V. Pavlov and A. Vasilev, 1972. L'effet de la saumure, de la congélation, et de la conservation de la viande de cheval sur la qualité des saucisses cuites. Mjasnaja Industrija SSSR 43: 27-28.

Pavlovskij, P. and V. Pavlov, 1979. Changements de l'activité protéolytique du tissu musculaire au cours de la maturation de la viande de cheval. Mjasnaja Industrija SSSA 2: 35-36.

Pitre, J., 1975. La viande. Connaissance biologique et bases de la technologie. Institut du Lait, Caen, France, 313 pp.

REFERences network, 2009-2013. Réseau economique de la filière equine en France, Haras Nationaux Editions, Paris, France, 40 pp.

Renerre, M., 1982. Influence de l'âge et du poids à l'abatage sur la couleur des viandes bovines (races Frisonne et Charolaise) Science Aliments 2: 17-30.

Robelin, J., Y. Geay and C. Beranger, 1977. Evolution de la composition corporelle des jeunes bvoins mâles entiers de race Limousine entre 9 et 19 mois. Composition anatomique. Annales de Zootechnie 26: 533-546.

Robelin, J., 1978. Répartition des dépôts adipeux chez les bovins selon l'état d'engraissement, le sexe et la race. Productions Animales 34: 31-34.

Robelin, J., R. Boccard, W. Martin-Rosset, M. Jussiaux and C. Trillaud-Geyl, 1984. Caractéristiques des carcasses et qualité des viandes. In: Arrige, R. and W. Martin-Rosset (eds.) Le cheval reproduction – sélection – alimentation – exploitation. INRA Editions, Versailles, France, pp. 601-610.

Rossier, E. and C. Berger, 1988. La viande de cheval, de qualités pourtant méconnues. Cahier de Nutrition et de Diététique 23: 35-40.

Roy, G. and B.L. Dumont, 1976. Système de description de la valeur hippophagique des équidés, animaux vivants et carcasses. Revue Médecine Vétérinaire 127: 1347-1368.

Salmi, A. and J. Hirn, 1981. Der cadmiumgehalt des muskels, der leber und nieren finnischer pferde und rentiere. Fleischwirtschaft 61: 1199-1201.

Trillaud-Geyl, C. and W. Martin-Rosset, 2011. Pasture practices for horse breeding. Synthesis of experimental results and recommendations. Fourrages 207: 225-230.

Tuleuov, E.T. and A. Billalova, 1972. Utilisation rationnelle de la viande de cheval. Mjasnaja Industrija SSSR 1: 30-31.

The new equine economy in the 21st century

Tuleuov, E.T., S. Tuleulova, A. Dikshtein and A. Juferov, 1974. Production de saucisson à partir de viande de cheval fraîche ou concervée 4 à 5 jours. Mjasnaja Industrija SSSR 1: 18-19.

Tuleuov, E.T., S.D. Ulyanof, S. Isenbaev, E. Bakman, A. Neeshlhkebof and A. Karpik, 1976. Effet du plasma sanguin sur la qualité des saucissons cuits de cheval. Mjasnaja Industrija SSSR 1: 14-16.

Ulyanof, S.D. and E.T. Tuleuov, 1976. Etude des changements autolytiques de la viande de cheval. In: European meeting of meat research workers-kavlinge. Swedish Meat Research Center, Malmo, Sweden.

Vermorel, M., J.P. Jouany, M. Eugène, D. Sauvant, J. Noblet and J.Y. Dourmad, 2008. Evaluation quantitative des émissions de méthane entérique par les animaux d'élevage en 2007 en France. INRA Productions Animales 21: 403-418.

Vervack, W., M. Vanbelle and M. Foulon, 1977. La teneur en acides aminés de la viande. Revue des fermentations et des industries alimentaires 32: 16-20.

Vial, C., G. Bigot, B. Morhain and W. Martin-Rosset, 2012. Territories and grassland exploitation by horses in France. In: Saastamoinen, M., M.J. Fradinho, S.A. Santos and N. Miraglia (eds.) Forages and grazing in horse nutrition. EAAP publication no. 132. Wageningen Academic Publishers, Wageningen, the Netherlands, pp. 467-480.

Part 9.

National reports – the horse in the national economy

Finally, this book includes a part containing National Reports on the state of horses. This is something of which, across Europe, there is a severe shortage. Reviewing the state of the art of research into horses in Europe, it is clear that this is something necessary in every state in the Union. It is, indeed, the hope of the authors of this book that such research can be supported by research funding in the not too distant future. We began the Introduction by claiming that there is a need for detailed, critical research into the numbers of horses in Europe, and into the economic impact equine activities generate there. These three chapters form a taste of what that research can reveal. Interestingly, the reports come from Iceland, Canada, and Russia – not the most mainstream of European equestrian nations! – and raise some interesting issues, particularly in their differences. Iceland has a very unique situation vis a vis horses with such high numbers of horse per capita, and this influences the way the equine economy works there in special ways. Canada, being such a large nation, seems to have an equine sector which is regionally fragmented. This makes it hard to characterize the Canadian equine sector and does cause some problems for riders across the nation. The final report, from Russia, reveals some very interesting aspects to the scene there, including the threatened continuation of a sizable horse meat sector, and the way in which equestrian sport has been concentrated under the Office of the President of the nation.

These reports then, are the result of careful research and can point the way towards a discussion of shared methodology which will inform further discussions along the way to undertaking proper national and international research on the growing role of horses in modern urban life, on the actual numbers and economic impact they have across Europe, and, from there, on policies which encourage both riding and horse welfare.

19. The new equine economy of Iceland

I. Sigurðardóttir[1]* and G. Helgadóttir[1,2]
[1]*Department of Rural Tourism, Holar University College, Holar, Iceland, 551 Sauðárkrókur, Iceland; inga@holar.is*
[2]*Tourism Management, Department of Economics and Informatics, Telemark University College, Gullbringvegen 36, 3800 Bø i Telemark, Norway*

Abstract

When many other horse breeds lost their role following the industrial revolution, the Icelandic horse gained a new economic role. This chapter explains the development of this new role of the Icelandic horse and what economic and social factors it includes. The research is based on a literature review and analyses of secondary data. It also collates information from the authors' prior research on equestrian tourism. Findings indicate that the new economic, cultural and social roles of the Icelandic horse are extensive and many-sided. It includes breeding in the country of origin, participation in events and shows, inclusion in various social activities in Iceland and abroad and multifarious business operations including e.g. breeding, training and equestrian tourism. Despite the fact that equestrianism is a lifestyle choice of its practitioners, the new equine economy in Iceland does not only rely on the domestic market, as international markets with Icelandic horses and services are substantial. The horse industry is in many ways similar in Iceland and other countries, but it is unique in that there is only one breed in the country. The Icelandic horse is highly recognised and popular worldwide. The breed has a worldwide studbook and an acknowledged country of origin. There is a very high number of horses per person in Iceland, a strong culture of equestrianism and expansive grassland available. This allows for horse husbandry that lets horses live according to their nature as herd animals, free ranging for part of the year and for young horses to grow up with mature horses with minimal human interference.

Keywords: equestrianism, horse industry, horse events, horse tourism, economic importance

Introduction

The aim of the chapter is to collate information from previous research to create an overview of the contemporary use, economic and cultural value of the Icelandic horse. This overview is based on a research program dating back a decade involving several projects. Quality in horse based tourism (Helgadóttir and Sigurðardóttir, 2008), the culture of horsemanship in Iceland (Helgadóttir, 2006), horse round-ups as a tourism attraction (Helgadóttir, 2015), native breeds tourism and a research on equestrian tourism as an industry in Iceland, including research on the demand side (Sigurðardóttir and Helgadóttir, 2015), the supply side (Sigurðardóttir, 2015) and cluster development (Sigurðardóttir and Steinþórsson, 2015).

The research methods range from qualitative fieldwork such as interviews and participant observation to visitor surveys and analysis of secondary data such as export statistics and available key figures on the horse economy. Published sources on the history of the Icelandic horse and relevant publications on Icelandic horsemanship were consulted, as well as the publications of stakeholders such as the international association of Icelandic horse breeders, the farmers' association and the ministry of Agriculture. The chapter examines the morphology of the horse sector in Iceland, rather than addressing economic evaluation. The focus is on describing the new economic role of the horse in a society that for ages depended highly on the horse in transport and work, cherished its riding abilities and has in the last decades created a new role for the horse exclusively as a leisure and sport horse.

We will first present the breed characteristics, provide a brief historical note to explain the claim for a pure breed and account for the breed numbers domestically and abroad. The scope of the horse industry in Iceland is outlined, insofar as statistical information allows. This is prefaced by a definition of a horse industry. Finally, we discuss central ideas about the link between breed and cultural/natural landscape in the equine economy, marketing discourses and contemporary horsemanship in Iceland.

The breed

The main distinction of the Icelandic horse is its gaits. The Icelandic horse is distinctive in that it has five gaits, as the horse can pace (skeið) and tölt in addition to walk, trot and gallop/canter (Björnsson and Sveinsson, 2006). This is central in the breeding, training and marketing of the breed. In the official breeding objectives emphasis is put on the light build, musculature and suppleness supporting the variety of gaits (Reglugerð um uppruna og ræktun íslenska hestsins 442/2011). This versatility makes it an interesting sports horse and is the basis for breeding, sports competitions and shows of the breed (Stefánsdóttir et al., 2014). It is also important in equestrian tourism as changing gaits according to terrain and to vary the pace and rhythm for travelling horse and rider, is important.

The Icelandic horse is also marketed for good temperament, riding ability and good looks. This is reflected in the official breeding objectives in items such as willingness, reliability, responsiveness and good temperament (Reglugerð um uppruna og ræktun íslenska hestsins 442/2011).This of course makes the breed popular in equestrian tourism and as a riding and sports horse in Iceland and abroad.

The Icelandic horse is rather small; the average height of a breeding horse was 138.5 cm at the withers in 2005, but the average height keeps increasing (Björnsson and Sveinsson, 2006). In a research on breeding horses published in 2014, the average height of breeding horses was 141±2.7 cm (Stefánsdóttir et al., 2014). Due to the small size, there has been some tendency to label the breed as a pony. 'Despite its small stature, the Icelandic Horse is referred to as a horse rather than a pony' (Edwards, 1993, p. 48). In the international community around the breeding, training and in tourism based on the Icelandic breed it is referred to as horse. A tongue in cheek video is out on the web explaining this to the world (The true Icelandic horse, n.d.).

In a research report on the colour and colour combinations of the Icelandic horse, the researchers showed that the eleven basic colours have a wide range of shades. With the methods they used approximately thirty shades were described and many different types of markings on head, body and legs have been identified. As the breed grows a winter coat there can furthermore be great variation in the colour of an individual depending on season (Þorvaldsson and Stefánsdóttir, 2008). This characteristic of colour diversity and that of a full mane, is an official breeding objective (Reglugerð um uppruna og ræktun íslenska hestsins 442/2011). This has made the colour diversity and coat an important selling point for the breed.

Horse farming in Iceland mostly means keeping horses in large herds and requires access to grazing land both on farms and in the North, on commons for weeks in the summer. The composition of the herd is based on feeding needs and breeding objectives. Breeding mares are kept with a stallion for the fertile period, but for most of the year foals, mares and growing youngsters (except stallion-prospects) graze together in larger groups. Stallions are kept in enclosures. An increasing number of riding horses spend their time in the peri-urban fringe of stables and smaller enclosures but the common practice is still to bring them to summer pasture for some weeks (Sigurjónsdóttir et al., 2003).

A note on the history of the Icelandic horse

The Icelandic horse has been from settlement and still is the only horse breed in the country. Settlers brought the Icelandic horse to Iceland from Norway and the British Isles in a period estimated between 860 and 935. In Iceland, it is exclusively pure bred, as there has been no import of other breeds for about 1000 years (Björnsson and Sveinsson, 2006). This is the foundation for claiming that the unique characteristics of the Icelandic horse are shaped by Icelandic nature.

From settlement to the industrial revolution, the horse was the most useful servant in the agricultural economy that underpinned Icelandic society. It was essential for travelling in rough landscape without a road system where rivers were unbridged and paths could lead through lava fields and across glaciers. Without roads, wagons were of no use so horses were for riding and as packhorses. While the horse was used for transport and briefly as a draught horse for farm equipment, it always retained its role as a riding horse (Björnsson and Sveinsson, 2006). It was a sign of status within the farming society to own good riding horses and specially to be able to keep a riding horse in a stable and feed it through the winter. Most horses had to strive through the winter by themselves, until the 20th century. Because of the lack of fences, breeding was not organized. The first sign of focused horse breeding was establishment of committees in every parish in the Skagafjörður region in North West of Iceland in 1879, to work on improvements in horse breeding. In 1891, the first law on horse breeding came into effect (Magnússon, 1978).

In 1915, George H.F. Schrader wrote a book on horses and equestrianism in Iceland. His goal was to describe to Icelanders how foreigners saw their horsemanship. He stated that the horse was the best Iceland had to offer. However, he was less than fascinated by the gaits the breed had. He stated that tölt and pace was common and preferred by Icelanders but in his opinion, these gaits were ugly and only suitable for second-rate riders (Schrader, 1986). Times have certainly changed in this regard.

While official agricultural consultancy dates back to around 1900 the first official consultant in horse breeding was appointed in 1920 (Magnússon, 1978). It took decades to agree on a breeding strategy for the Icelandic horse. There was always a strong focus on tölt and pace, despite the suggestions of Schrader and others who argued against it. There was also a disagreement on whether to breed two types of horses; a working horse and a riding horse. The two-type strategy did not gain ground, as before it could be implemented the prospect of mechanical power replacing horses as beasts of burden became evident. Since mid-twentieth century, the breeding has focussed on riding ability as the role of the horse is increasingly in recreation rather than farming (Jónasson, n.d).

The current regulations on the breeding aims of the Icelandic horse include a statement on Iceland being the country of origin for the breed, regulations on registration of individual horses aims regarding health and fertility, as well as aims for evaluation of conformation and riding abilities (Reglugerð um uppruna og ræktun íslenska hestsins, 442/2011). The international association for the breeding and training of the Icelandic horse honours this in their regulations (International Federation of Icelandic Horse Associations, n.d.).

Equestrianism is today a widely practiced sport and popular recreation in Iceland. On the list of The National Olympic and Sports Association of Iceland, of most practiced sports in the country the sport of equestrianism has been in third place, with soccer in first place and golf in second. According to sport statistics, 10,214 Icelanders practiced equestrianism in 2008 (http://tinyurl.com/p8svvxe). Those numbers represent the members of the Icelandic Equestrian Association. Not everyone taking part in equestrian sport in Iceland is a member of the association, but the estimated number of Iceland residents, doing so in some manner is 35-40,000 (Stefánsdóttir, 2009) out of 329,000 inhabitants.

The new equine economy in Iceland is rooted in the common history of horses and humans in Iceland and the cultural importance of the horse (Helgadóttir, 2006). The focussed breeding over the last decades has concentrated on a purebred riding horse. This is the foundation of an equine economy where the horse is no longer the most useful servant but a valued co-worker in a new variety of businesses, events and recreational activity in Iceland and abroad. While it is possible to speak of a new equine economy in Iceland, it is important to note that horse farming and husbandry is still based on age old traditions.

The number of Icelandic horses

In 2010 the registered number of horses in Iceland was 77,158 (http://tinyurl.com/o4846fc) so there were 240 horses per 1000 persons in Iceland (Sigurðardóttir, 2011), a much higher ratio than in other European countries. According to a research report on horses and the horse industry in Europe, published in 2009, the largest population of horses per 1000 inhabitants in European Union countries is 31 horses in Sweden. The average number of horses per 1000 persons, in the 23 European countries listed in this report, was 13 (Liljenstolpe, 2009). Iceland was not included in this research as it is not part of the European Union.

Available data from Statistics Iceland (2014) show that over the last 30 years the numbers have been fairly stable, between 70,000 and 80,000 horses registered in Iceland. According to Statistics Iceland, the number of horses increased in Iceland from 52,000 in 1984 to 80,500 in 1996. After that the number went down until 2002 when it started rising again. In 2002 there were 71,012 horses registered in Iceland, compared to 77,380 in 2012 (Icelandic Food and Veterinary Authority, 2014). Therefore, the horse population in Iceland was at that point growing.

According to a farm structure survey conducted by Statistics Iceland, in 2010 there were 1,338 farms in Iceland that had 10 or more horses (Statistics Iceland, 2012). The North West of Iceland has 2.6 horses per person (Figure 1) which is the highest ratio of horses per person in the country (Sigurðardóttir, 2011). It has been estimated that 212 horse breeders operate in that area along with a considerable engagement in other kind of horse based activity and businesses from all levels of the value chain (Sigurðardóttir and Steinþórsson, 2015).

The horse population in Iceland is however only part of the total number of Icelandic horses, as Icelandic horses have been exported for decades. About 100,000 Icelandic horses are registered abroad. The total number of registered foals born overseas in 2013 was 10,614, as opposed to 4972 in Iceland (International Federation of the Icelandic Horse Associations, 2014). This shows that the breed is firmly established overseas.

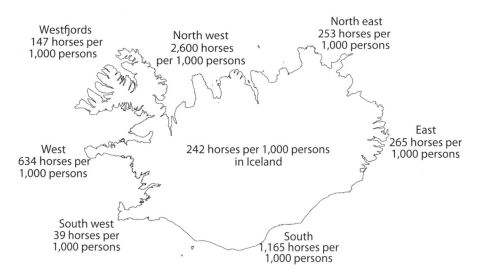

Figure 1. The number and distribution of horses and residents in Iceland.

The international studbook of the breed, 'World Fengur' is important for the development of the horse breed. It also supports the foundation for the marketing of Iceland as the country of origin of the Icelandic breed by documenting how the bloodlines lead to the country of origin. The International Federation of Icelandic Horse Associations (FEIF) currently has 59,157 members in 19 countries. It has 500 clubs in the member countries, there of 45 in Iceland (International Federation of Icelandic Horse Associations, 2014).

An international target market thus exists for the Icelandic horse as it is fairly well known and its characteristics are consistently promoted. This creates marketing opportunities for Icelandic equestrian businesses. This is an export market for riding, breeding and competing horses as well as for various horse related services. Training, riding instruction, shoeing, judging and breeding are among the services that the Icelandic equestrian sector exports. Icelandic specialists emigrate, commute and import clients to Iceland for horse business. Holar University College has a bachelor programme in riding, training and riding instruction based on and tailored to the Icelandic breed attended by both the Icelandic and international students (Holar University College, n.d.).

What is a 'horse industry'?

In our research, we define the horse industry as all activity based on and/or related to horses and equestrianism, including core activities as breeding and training as well as related activities and services. Examination of academic writings of horsemanship/equestrianism as an industry indicates lack of research and writings of the topic. However, a few characteristics of the industry are consistently documented in the literature; diversity, fragmentation and lack of reliable statistical information on the sector.

In 2004 The Henley Centre brought forward a definition of the British horse industry where it defines 'the horse industry as encompassing all activity that has the horse as its focus and activity that, in some reasonable capacity, caters for such an industry' (2004: 9). The U.S. horse industry is a very large and important part of national, state and local economies. The industry is 'diverse, involving agriculture, business, sport, gaming entertainment and recreation' (The American Horse Council Foundation, 2005).

Globally the horse industry is fairly diverse and fragmented, with only the horse as a common denominator. It is reasonable to assume that the Icelandic horse industry is also diverse and possibly fragmented, but it has not been a focus of much research as an industry. Some segments of Icelandic horsemanship/equestrianism are defined as agriculture, but others as leisure activity of horse owners and guests in equestrian tourism. As for fragmentation, it should be noted that as there is only one horse breed in the country this forms as a very strong common foundation for all aspects of the industry. There are some indications of less specialization than in other countries. That is industry actors having multiple roles for instance in breeding, training, competing and shoeing horses (Sigurðardóttir and Steinþórsson, 2015).

A definition of equestrianism as an industry has, however, not been agreed upon in Iceland. In Iceland, as abroad, riding has been one of the uses for horses and riding has encompassed many things such as transportation, work, sport, art and leisure activity (Björnsson and Sveinsson, 2006; Helgadóttir, 2006).

Characteristics of the horse industry

Businesses within the horse sector are in most cases small and medium sized. The horse industry in Europe seems to have strong traditions and a strong base in 'traditional' agriculture and heritage, resulting in conservative ('old boys club') thinking with limited focus on markets and customers

(G. Dalin, personal communication). Unprofitable activity, hard work (physically and mentally) and health problems are listed as drawbacks of the industry in Finland (T. Thuneberg, personal communication). In Britain 'Equestrian businesses tend to be run on lower than average profit margins than comparable businesses in other sectors, and have little awareness of how they are performing relative to others' (British Horse Industry Confederation, 2005:19).

The problem of low profit from the horse sector has been a longstanding concern in Iceland (Hagþjónusta landbúnaðarins, 1998). Findings from a 2008 Icelandic research project among operators in horse-based tourism suggest their disinterest in economic value or the business side of the enterprises. The 40 operators interviewed turned out to be more interested in and more knowledgeable about breeding, training and riding of horses than in business operation, including finances, insurance and business plans (Helgadóttir and Sigurðardóttir, 2008). Those findings comply with findings from Britain; horse-based businesses are run by people who are more interested in horses than in business (British Horse Industry Confederation, 2005). That is horse-based tourism is a life-style entrepreneurship (Anderson Cederholm and Hultman, 2010).

Research indicates that in some sections of the horse industry in Europe, self-appointed 'experts' appear to be common, and management seems to be in some cases done by belief rather than rational study, so there is a need for 'a leap forward' (G. Dalin, personal communication). Developing businesses, i.e. by making business and marketing plans, is important, as the businesses in some cases seem to be 'non-focused' to the point where the managers cannot name their main sector of activity (The Henley Centre, 2004; T. Thuneberg, personal communication). For the last 10 years, development of the horse industry in Europe seems to have been oriented towards diversification rather than growth (Liljenstolpe, 2009). Growth and financial profit have shown not to be the most important indicators of business success, in the minds of operators in equestrian tourism. 'Lifestyle components like making a personal hobby financially manageable, creating jobs for the family and increasing the utility of their resources like land, horses, buildings and equipment seem to be important' (Sigurðardóttir, 2015). Demand for increased quality in various sectors of the industry is apparent (Helgadóttir and Sigurðardóttir, 2008; The Henley Centre, 2004; H. van Tartwijk, personal communication).

Research does indicate that equestrian businesses commonly have the characteristics of lifestyle enterprises (Helgadóttir, 2006; Helgadóttir and Sigurðardóttir, 2008; Anderson et al., 2010; Anderson Cederholm, 2012; Sigurðardóttir, 2015). Helgadóttir (2006) points out that horsemanship can include business operation, hobby and lifestyle. Liljenstolpe (2009) states that horses can represent a leisure or sporting activity, a way of life, a working companion or food consumption. In a discussion of horse keeping in the urban-rural fringe, Elgåker and Wilton (2008:45) state that 'projects and research about development and innovation in rural areas often examine new types of enterprise in rural areas, but small land-holdings where lifestyle and recreation is combined is often forgotten since each unit is often small and not meant for economic production'.

The horse sector differs from other agrarian sectors in that agricultural policies that ensure production premium for farmers do not extend to the horse sector. In Iceland, dairy production and meat production from sheep and cattle receives state support based on production while breeding horses does not. This may also contribute to the big discrepancy in documentation of the extent of the industry in Iceland as elsewhere, as documentation is not a condition for income as in those sectors receiving public funds.

Tourism and horses in Iceland

The terminology on and definitions of tourism involving horses is diverse as pointed out by Buchman (2014), who advocates for using the most inclusive term 'horse tourism' suggested by the Equine research network. Horse tourism in Iceland is an important niche in Icelandic tourism, which partly relies on the target market of the Icelandic horse enthusiasts abroad who may wish to experience the horse in its native environment. The Icelandic horse is a part of Iceland's image as a nature based tourism destination. About 80% of all visitors state that their interest in the Icelandic nature affected their decision to visit Iceland (Óladóttir, 2013). The Icelandic horse is known worldwide not only among horse enthusiasts but also among those that look to Iceland as a destination for leisure travel and recreation.

Visitor surveys taken at points of departure from Iceland show that of overseas visitors between 15-20% go horseback riding while visiting Iceland, the percentage during high season is higher but the numbers during low season are down to 15% (Óladóttir, 2004; Óladóttir, 2005; Helgadóttir and Sigurðardóttir, 2008). Thus in 2014, about 160,000 foreign guests did purchase short or long horse riding tours. In 2002 the estimated number was 42,000 (Sigurðardóttir, 2005). This may reflect a global trend in that horse tourism, both equine and equestrian, is a growing niche in world tourism (Ollenburg, 2005; Buckley *et al.*, 2008; Boyer, 2012; Vaugeois, 2012; Buchman, 2014). Despite the estimation of the number of foreign guests in riding tourism in Iceland, figures are still lacking on the price and length of the tours and the ratio of guests buying short vs long tours. This means that estimating the total turnover of the industry is hard.

Horse tourism in Iceland consists of variable activities. Most commonly short and long riding tours (Helgadóttir and Sigurðardóttir, 2008), but increasingly of horse shows, horse theatre and farm stays ('Hestatengd ferðaþjónusta í blóma', 2014). It has also been pointed out that opportunities for further product development within the horse tourism field can be identified within the field of wellness tourism (Sigurðardóttir, 2014). Some events within the Icelandic equestrianism are however, more focused on domestic participants and spectators than international tourists. Numbers of guests in types of horse based tourism other than short and long riding tours is unknown.

The number of domestic equestrian tourists is more difficult to estimate. In a nationwide survey on travel behaviour of domestic tourists, 24 out of 970, that is to say 3.2% of respondents, had paid for a horse ride during their vacation that year (Markaðs- og miðlarannsóknir, 2011). It must however be kept in mind that travelling with/on own horses for several days during the summertime is a common leisure activity among Icelandic horse owners and their leisure riding is not measured by this survey. It is likely that most Icelanders who travel by horse or ride for leisure, do ride their own horses or borrow horses from friends or relatives. Those groups of travellers do not buy the same kind of service as most guests of the horse based tourism businesses. However, they do buy various kinds of services during their travel, like accommodation for the travelling group and a pasture space for up to 100 horses per group (Schmudde, 2015). Information on the number of travellers, length of tours and average spending during those tours is not available, but would be an interesting subject of a future research.

Horse events in Iceland

Events are an important aspect of every industry, both for promotion of products and services and as a celebration of the industry subculture. The main horse events in Iceland have traditionally fallen into four categories; breeding shows, sports events, horse shows or horse theatre and horse gathering/round-up. The contemporary horse sector has a range of events, many of them dating back centuries, such as horse gatherings or roundups, horse markets, horse shows, races and equestrian sports.

Breeding shows

The purpose of the breeding shows is to evaluate and rank potential breeding horses and later on to calculate a best linear unbiased prediction (BLUP) for individual breeding horses. These shows are the forum for development in the breeding, a market place and show case for the horse industry.

Evaluation of breeding horses takes place during the summer time. In a breeding show, horses are first assessed for conformation, then for riding ability. Horse breeders often hire top riders and trainers to prepare and ride the horses in the show.

There are detailed regulations for breeding shows both pertaining to horses, riders, the staff of the show and the facilities. Judges in these shows must conform to the industry standards, regulations and have a university degree in agriculture as well as having passed a qualifying exam and/or been trained on courses approved by FEIF. Three judges must be present and the show must have an event manager (Reglugerð um uppruna og ræktun íslenska hestsins, 442/2011).

Sports events and competitions

Sport events and competitions are discussed together as they have a similar format and attract a common target market; domestic and international breeders, trainers, recreational riders and horse and sports enthusiasts. Competing in the events are top horses and riders of each time, leisure riders and young riders. They typically also consist of entertainment related to horses and horse culture.

The sports category involves various gaited competitions. Horses are assessed for versatility, control, reliability and rhythm in the gait, grace of movement and beauty in riding. The shows are indoors in winter, with the exception of outdoor ice tölt while summer events are outdoors. The horse clubs, that is the sports practitioners have local and regional competitions starting in January-February and leading up to a couple of culminating events for practitioners of all levels and domestic and international spectators in the spring and summer.

Races are not an extensive part of horse events in Iceland, but races in gallop and flying pace exist. There is more emphasis on flying pace. Betting on the winner is not part of the culture.

The national championship of the Icelandic horse, the Landsmót has been held since 1950 and from 2000 as a biannual event. It is the national championship, as well as the biggest festival and industry event for the Icelandic horse sector. The Landsmót is also a tourist attraction. Landsmót is a week-long event, which includes breeding shows, gaited competition and shows as well as evening entertainment. It is an outdoor event, camping is arranged on or near the site and supported by a range of services such as catering, sales of horse related products and services as well as children's playground. The first Landsmót in 1950 was attended by 10,000 guests and the Landsmót in 2008 by 14,000. Attendants come from most of the countries that have Icelandic horse populations (https://landsmot.squarespace.com/about-landsmot/).

The Landsmót has a counterpart in the World Championships of the Icelandic horse, which is held on the alternate year outside Iceland. The World Championships are the second major event, but for the aforementioned quarantine reasons they cannot be held in Iceland. Horses from Iceland that compete in the World Championships can for the same reason not return (Reglugerð um uppruna og ræktun íslenska hestsins nr. 442/2011).

Horse entertainment

Horse businesses, especially horse tourism businesses organize horse shows to introduce the Icelandic horse, its gaits, colours and the history of the horse. Such shows are also part of the entertainment during major sport competitions. They consist of a combination performance of gaits and dressage elements.

In the summer the shows usually take place outdoors. One of those shows has attracted 70,000 guests in 7 years and has grown from 900 visitors the first year (Hestatengd ferðaþjónusta í blóma, 2014).

The newest example of the development of horse shows within the tourism industry is the horse theatre Fákasel which offers an introduction of the Icelandic horse, in the form of a 'command performance'. Fákasel opened in February 2011 and regularly offers a show named 'The Legend of Sleipnir' (Hestatengd ferðaþjónusta í blóma, 2014).

Horse gathering / round-ups

This refers to the practice of gathering domestic horses from the highland grazing commons, herding them to a lowland fold or corral, and then sorting and bringing them home to their respective farms for the winter. This is also a practice in sheep farming and the same pastures, routes, enclosures, fences and corrals are used for horses and sheep (Helgadóttir, 2015).

In North West Iceland the mares and their foals as well as growing youngsters are herded to mountain pastures in the summer. In late September, early October they are gathered. It takes from a day, to up to three days of riding to gather the horses. The original purpose of the gathering is to retrieve the horses. The horse owners look forward to seeing how their foals have grown in summer pasture (Helgadóttir, 2006; Helgadóttir, 2015).

These events are marketed both as equestrian tourism and as local festivals. The ratio of non-resident individuals at the horse roundups is over 70%, mostly domestic tourists as the percentage of international tourists' only ranges from 6 to 17% of visitors (Helgadóttir and Sturlaugsdóttir, 2009). The international tourists typically buy a package tour of about a week from a tour operator that arranges travel, accommodation, meals and entertainment. The operator also arranges horses, pre-gathering rides and guides to lead the tourists during the gathering and roundup.

Those events are authentic from the farmer's point of view, as they are first and foremost, about gathering and sorting their horses. They have also provided tourism operators with the opportunity to extend the high season in their tourism services. All accommodation in the neighbourhood of each round up is fully booked months ahead, catering services are busy during the weekend and all horses available to tourists in the region are in use at the day of the round up.

Conclusion

The horse sector in Iceland is as in other countries quite diverse but it is questionable if it is as fragmented as elsewhere. It is a proposition which needs further research whether the fact that there is only one breed of horses, leads to cohesion rather than fragmentation of the horse industry in Iceland. It is a lifestyle entrepreneurship sector with the common characteristics of making a living from one's passion rather than making money from it. While horse people around the world may think somewhat alike there are a few unique local characteristics of the equine economy in Iceland.

It is unique to have only one breed of horses in a country. This has been successfully utilised in marketing through a formal designation of country of origin, so that Icelandic horses have a 'denomination d'appellation' if you will. There is one studbook for the breed (WorldFengur), which again is unusual and a strong marketing point. The export of Icelandic horses and the international organisation around the breeding, training, showing, competing and leisure riding and education in horsemanship through FEIF forms a solid base for the sector.

The decision taken in the early twentieth century to focus breeding and training on riding qualities, particularly the five gaits creates an advantage for the breed in today's leisure horse economy. In this regard, it is also important that due to historical development of the transport infrastructure, the Icelandic horse was not traditionally a draught horse but always a riding horse.

The high number of horses in the country and the adherence to traditional ways of keeping horses as the herd animals they are by nature makes the culture of horsemanship an important equestrian heritage. With good access to land, it is possible to keep horses as the herd animals they are and allows foals to be raised by other horses with minimal human interference. The practice of keeping horses outside year round ensures that they have space for movement and that they retain the characteristic of growing a thick winter coat.

Last but not least, the above mentioned characteristics have led to a market opportunity in equestrian tourism that is well used. The percentage of international tourists that ride in Iceland is exceptionally high and the numbers of domestic recreational riders are very high compared to other European nations. Consequently, it is not surprising that the profile of the horse in marketing Iceland is high and reflects its iconic status in contemporary culture.

References

Andersson Cederholm, E., 2012. Lifestyle enterprising: the ambiguity work of Swedish female horse-farmers. Research in Service Studies, Working paper 10: 1-20.
Andersson Cederholm, E. and J. Hultman, 2010. The value of intimacy-negotiating commercial relationships in lifestyle entrepreneurship. Scandinavian Journal of Hospitality and Tourism 10(1): 16-32.
Björnsson, G.B. and H.J. Sveinsson, 2006. Íslenski hesturinn. Mál og menning, Reykjavík, Iceland.
Boyer, C., 2012. The infrastructure of equestrian tourism. In: Conference proceedings equi-meeting tourisme, cheval, tourisme et loisirs, transformations, permanences et ruptures. Saumur, France, pp. 167-168.
British Horse Industry Confederation, 2005. Strategy for the horse industry in England and Wales. Available at: http://tinyurl.com/c2obubj.
Buchman, A. 2014. Insights into domestic horse tourism: the case study of lake Macquarie, NSW, Australia. Current Issues in Tourism.
Buckley, R., C. Ollenburg and L. Zhong, 2008. Cultural landscape in Mongolian tourism. Annals of Tourism Research 35(1): 47-61.
Edwards, E.H., 1993. Horses. A Dorling Kindersley Book, London, United Kingdom.
Elgåker, H. and B.L. Wilton, 2008. Horse farms as a factor for development and innovation in the urban-rural fringe with examples from Europe and Northern America. In: Proceedings from 10th Annual Conference, Nordic-Scottish University for Rural and Regional Development. University of Copenhagen, Copenhagen, Denmark.
Helgadóttir, G. and R. Sturlaugsdóttir, 2009. Stóðréttir: avinningur af komu ferðafólks. Ráðstefna um íslenska þjóðfélagsfræði, The University of Akureyri, Akureyri, Iceland.
Hagþjónusta landbúnaðarins, 1998. Hrossabúskapur og hrossaeign á Íslandi 1996, úttekt og stöðumat. Hagþjónusta landbúnaðarins, Hvanneyri.
Helgadóttir, G., 2006. The culture of horsemanship and tourism in Iceland. Current Issues in Tourism 9(6): 535-548.
Helgadóttir, G., 2015. Horse round-ups: harvest festival and/or tourism magnet. Cheval, Tourisme & Sociétés/Horse, Tourism & Societies, Mondes du Tourisme, special issue – hors série, June: 216-223.
Helgadóttir, G. and I. Sigurðardóttir, 2008. Tourism: community, quality and disinterest in economic value. Scandinavian Journal of Hospitality and Tourism 8(2): 105-121.
Hestatengd ferðaþjónusta í blóma, 2014. Eiðfaxi 9(38): 6-9.

Holar University College, e.d. Department of Equine Studies. Available at: www.holar.is/en/department_of_equine_studies.

Icelandic Food and Veterinary Authority, 2014. Available at: http://tinyurl.com/o4846fc.

International Federation of Icelandic Horse Associations (FEIF), 2014. Statistics about member countries: registered horses. Available at: www.feif.org/FEIF/Factsandfigures.aspx.

Jónasson, J., (n.d.). Landbúnaður á Íslandi. Bændasamtök Íslands. Available at: http://tinyurl.com/p4ytrjx.

Liljenstolpe, C., 2009. Horses in Europe. Swedish University of Agricultural Sciences, Uppsala, Sweden.

Magnússon, S.A., 1978. FÁKAR, Íslenski hesturinn í blíðu og stríðu. Bókaforlagið Saga, Reykjavík, Iceland.

Markaðs- og miðlarannsóknir, 2011. Ferðalög Íslendinga. The Icelandic Tourism Board, Reykjavík, Iceland.

Ollenburg, C., 2005. Research Note, worldwide structure of the equestrian tourism sector. Journal of Ecotourism 4(1): 1-9.

Óladóttir, O.Þ., 2004. Könnun meðal erlendra ferðamanna sumarið 2004. Icelandic Tourism Board, Reykjavík, Iceland. Available at: http://tinyurl.com/ol67geh.

Óladóttir, O.Þ., 2005. Könnun meðal erlendra ferðamanna veturinn 2004-2005. Icelandic Tourism Board, Reykjavík, Iceland. Available at: http://tinyurl.com/nhbj56o.

Óladóttir, O.Þ., 2013. Tourism in Iceland in figures, April 2012. The Icelandic Tourism Board, Reykjavík, Iceland. Available at: http://tinyurl.com/owuk63m.

Reglugerð um uppruna og ræktun íslenska hestsins 442/2011. Available at: http://tinyurl.com/nh692uz.

Schmudde, R., 2015. Equestrian tourism in national parks and protected areas in Iceland. An analysis of the environmental and social impacts. Scandinavian Journal of Hospitality and Tourism 15(1-2): 91-104.

Schrader, G.H.F., 1986. Hestar og reiðmenn á Íslandi. Bókaútgáfan Hildur. Reykjavík, Iceland.

Sigurðardóttir, I. and G. Helgadóttir, 2015. Riding high: quality and customer satisfaction in equestrian tourism in Iceland. Scandinavian Journal of Hospitality and Tourism 15: 1-2.

Sigurðardóttir, I. and R.S. Steinþórsson, 2015. Hestar og þróun klasa; hestatengdur klasi á Norðurlandi vestra. Skrína 2(6): 1-9.

Sigurðardóttir, I., 2014. Equestrian tourism as a contributor to wellbeing and happiness. In: ATLAS annual conference 2014, tourism travel and leisure – sources of wellbeing, happiness and quality of life. Association for Tourism and Leisure Education and Research, Budapest, Hungary, pp. 39-43.

Sigurðardóttir, I., 2011. Economic importance of the horse industry in Northwest Iceland: a case in point. Economics, Management and Tourism, Second International Conference for PhD Candidates. South-West University 'Neofit Rilsky', Bulgaria, pp. 113-117.

Sigurðardóttir, I., 2005. Hestatengd ferðaþjónusta á Íslandi: atvinnugrein eða tómstundagaman? In: Jónsdóttir, R.S. (ed.) Fræðaþing landbúnaðarins 2005, I-Erindi. Fræðaþing Landbúnaðarins, Reykjavík, Iceland.

Sigurðardóttir, I., 2015. Identifying the success criteria of Icelandic horse-based tourism businesses: interviews with operators. Cheval, Tourisme & Sociétés/Horse, Tourism & Societies, Mondes du Tourisme, special issue – hors série, June: 150-160.

Sigurjónsdóttir, H., M.C. Van Dierendonck, S. Snorrason and A.G. Þórhallsdóttir, 2003. Social relations in a group of horses without a mature stallion. Behaviour 140: 783-804.

Statistics Iceland, 2014. Búfjártölur 1981-2013. Available at: http://tinyurl.com/o4846fc.

Statistics Iceland, 2012. Statistical series, fisheries and agriculture 2010-2012. Available at: http://tinyurl.com/o3mdzz7.

Stefánsdóttir, G., 2009. Hlutverk íslenska hestsins í skóla og samfélagi. Unpublished thesis, University of Akureyri, Akureyri, Iceland.

Stefánsdóttir, G.J., S. Ragnarsson, V. Gunnarsson and A. Janson, 2014. Physiological response to a breed evaluation field test in Icelandic horses. Animal 8(3): 431-439.

The American Horse Council Foundation, 2005. National economic impact of the U.S. horse industry. Available at: www.horsecouncil.org/nationaleconomics.php.

The Henley Centre, 2004. A report of research on the horse industry in Great Britain. Defra Publications, London, United Kingdom.

True Icelandic horse, n.d. Available at: https://vimeo.com/113323901.

Vaugeois, N., 2012. Equestrian tourism in British Columbia, Canada: its evolutions, current state and potential. Conference proceedings. Equi-meeting tourisme, cheval, tourisme et loisirs, transformations, permanences et ruptures. Saumur, France, pp. 22-24.

Þorvaldsson, G. and G.J., Stefánsdóttir, 2008. Litir og litbrigði íslenska hestsins. Rit LbHí 16. Landbúnaðarháskóli Íslands, Hvanneyri, Iceland.

20. State of the equine industry in Canada

H. Sansom
University of Guelph, School of Environmental Design and Rural Development, 50 Stone Road East, Guelph, ON, N1G 2W1, Canada; heather@heathersansom.ca

Abstract

The equine industry in Canada is diverse and complex, in part due to the geographical scale of the country of Canada, as well as the wide variety of equine disciplines and activities that are popular in Canada. While equine sports and activities remain popular in Canada, there are signs that the demographics are shifting to an older participant base, raising questions regarding the long-term sustainability of the industry. The distribution of the population in Canada also contributes to challenges in terms of developing and sustaining an active base for equine sports and horse-keeping in Canada. This leads to localized regions of equine activity making it difficult to share resources and to build national programs. This chapter provides an overview of the equine industry in Canada and current challenges facing the industry, from the perspective of industry participation and sport management. This chapter's informal style is due to its being a near-transcription of a half-hour industry overview presentation made by the author at the European Federation of Animal Science Conference (EAAP) in Copenhagen, Denmark in August 2014.

Keywords: equine industry Canada, Canadian equestrian, challenges, strengths

Introduction

There is a notable lack of academic material and official, publicly available organizational documentation of the equine industry in Canada. Much material for this discussion was obtained informally, either during the author's time in a management role with the National equestrian federation (Equine Canada), or through personal interviews and contact with individuals with access to non-published primary but un-published information. As a result, some information sources are not quotable since anonymity was preferred. The Equestrian Federation in Canada does not keep the more extensive records which are common in some European countries. The most recent industry report commissioned by Equine Canada dates to 2010 (Evans, 2011).

Despite gaps in documentation and academic study, sufficient information is available to provide some basic outlines of industry status and trends. The size of the country, number of people involved, number of organizations involved and diversity of viewpoints and stakeholders testifies to some major themes shaping the industry: the vast geography of Canada, changing demographics, wide diversity in equine practice and culture-both regional and by discipline, organizational silos and splinters, low popular engagement, and long historical roots. There are also some recent hopeful opportunities for new energy and focus, such as the upcoming PanAm Games and World Equestrian Games.

Organizational overview

The National equestrian federation, Equine Canada, is run by over 70 committees with volunteers from industry who are distributed across the country. The volunteers range from subject matter experts, to industry leaders, to interested parties with time and resources to be available. While Equine Canada serves as a lead organization for the country, it does not do so in the same capacity as European federations. Main areas of governance are policy, coaching and rider curriculum, competitive sport, advocacy with regard to government policy impacting equine activities, and providing a loose forum of association for all the other organizations. The larger percentage of the day-to-day organization of equine activity is conducted by Provincial, regional, discipline and breed associations which are affiliated with Equine Canada, leading obviously to wide diversity and minimum cohesion. Governance and management of other aspects of the industry is not kept by the federation, but by the other organizations involved. Examples are racing, meat, non-FEI discipline and breed organizations.

Industry overview

It would not be possible to get a strong picture of the industry without a basic understanding of the Canadian context. At the time of last census the Canadian population was estimated at 35,160,000. On average, 400,000 new members of the population are added yearly. Approximately 66% of new citizens are the result of international immigration. We actually otherwise have negative population growth, partly due to aging demographic, and partly trends in lower birth rates in middle class industrialized culture. It is estimated that approximately 20% or seven million Canadians are born outside the country. The number of Canadians whose parents were born outside the country is of course much larger than that. Canada is a country of immigration (Statistics Canada, 2011).

At the start, we were settled by French socio-cultural refugees and opportunists. It's no secret that Europe was at the time exploring for new resources to exploit, and for places to send non-fitting populations. The New World was a timely discovery. The French were followed by the British and we had waves of British settlers: religious and economic refugees, capitalists and adventurers. This was followed by Chinese who were not mentioned in history books until more recent years, and mostly Europeans – again political, religious and economic refugees and entrepreneurs for the most part. Settlement of the vast wild and forested continent by people who could fell trees, farm and build industries was important.

After the main European waves, there have been increasing waves of immigration with higher percentages from populations noticeably non-European, and largely again political, economic, religious refugees or those seeking a better life for self or children, just like the Europeans before them. The basis of equine activity in Canada comes from the times of the European settlers, and the face of the industry is still largely Caucasian and from relative affluence. In Canada, affluent non-Caucasian populations do not seem to be strong participants. This may be partly due to marketing of the industry which is largely by the traditional Caucasian horse-involved demographic, to their own demographic. There is little promotion of equine activity outside traditional participant bodies, to attract newcomers.

Affluence discussed is relative, with many leisure horse participants coming from income brackets where a choice is made between their horse activity, and other middle class conspicuous spending patterns such as eating out or taking vacations (Evans, 2011; Statistics Canada, 2011). As such, it is increasingly out of synch with the urbanized, multi-cultural, dominant culture, and also with the middle class majority which have come from working-class parents and worked their way into what is considered middle class in a consumerist culture (largely defined by income and spending patterns shaped by urban pop-culture, not so much by education, culture and vocation). Large numbers of horses in Canada does not equal the robustness of industry that one might think, when we consider the impact on horse-keeping of horse activity being practiced by an increasingly minor part of the general culture. Current numbers and future sustainability are two separate things.

Demographics: age, immigration, multi-culturalism, rural-urban shift

When we are talking about a sport which has been traditionally practiced by European cultures, the impact of immigration on sport culture is enormous. Sport practice in Canada is undergoing change. In the past two decades we have seen a significant shift from popular identification with hockey for example, to increased participation and media coverage of football (soccer). The traditional cultural home of equine activity has been in the habit of internal communication, and is not so practiced at communication and recruitment to the sport from other cultures. In the changing Canadian demographic situation, lack of attraction of newcomers to equine activity results in increased marginalization of equine activity from mainstream culture. It also ensures a strong dependence of the industry on the aging Baby Boom demographic who will soon be reaching ages of diminished equine related spending. By cultures, one could mean ethnic, but also other sport cultures, age-group cultures or even urban vs rural.

Aging population affects sport choice and personal resources available for involvement with horses. In Canada 16% of the population or more are in the 65 year and over age category. Projections for 2050 envisage up to 25% of the population in that category (Statistics Canada, 2011). To understand an industry, you have to understand its market needs and priorities. It is clear that there will be a severe impact from a mass exodus of participants into an age category where land ownership and physical activity around horses may have to be given up. There is a very well-known book called *Boom, Bust & Echo* (Foot and Stoffman, 1996) which was a big hit in the late '90's early 2000's when I was working on my MBA. The title summarizes the main point very well, and the idea has been around in North American economic and business discussion for some time. The general thesis of this book is that business needs to consider its relationship to the phenomenon of the post-World-War II aging 'baby boom' generation. Business success in North America in many cases is tied to a product or service applicability to the typical needs of this aging cohort.

It is difficult to be precise about a rural urban split in riding and horse ownership, particularly because of differences in definitions of rurality. The Canadian government defines rural areas in its census as those municipal areas with a population density of 10,000 or fewer. This includes small towns with 3,000 to 8,000 people in that town and in surrounding farm areas. The rural/urban population split in Canada has been declining over the years. In 1850 rural residents numbered 87%. After the First World War, by 1921 the number had dropped to 51%. It dropped again after the Second World War by 1961 to 30%. By 1981, the percentage of rural residents was 24%. A decade after that, in 1991, it was down to only 23%.

Currently at 19%, there now seems to be some stability (Statistics Canada, 2011). In many countries, a municipal population as high as 10,000 would even be called urban. Certainly, even within Canada what is considered rural is often different, depending on the purpose, subject, or speaker. For example people living in the North do not consider most parts of Southern Ontario rural. However, those in Southern Ontario with farms or who would live near fields or at driving distance to shopping would consider themselves rural.

In Canada, foot, bicycle or public transit access to shopping for basic necessities is very uncommon outside urban areas. Sidewalks or paths to connect countryside dwellers to basic amenities are virtually non-existent. Rights of way and facilitation of active transport in the countryside are not aspects of the culture or infrastructure. Most property is fenced, there are no public rights of passage or access, and roadways presume and provide for automobile traffic only except in very small pockets such as near an Amish area. The lived experience of countryside dwellers is such that rurality is defined by many in terms of whether or not a car is needed for survival.

There are very few urban horse facilities, which means that most horses are considered as housed in rural areas, both by popular definitions, and by Statistics Canada (2011). The distance between areas considered urban/suburban and those considered rural, and so the distance between the culturally dominant (urban) and horses, is getting wider. Lack of deeply historic urban areas, relative rapidity of development and support of the oil and auto industries has led to wide urban/suburban sprawl with less consideration for population access to natural and rural areas or for sustainability (Brodhagen, 2009; Burns *et al.*, 2009; Cloke *et al.*, 2006). As a result, horses are moved further away from the population (or the population has moved further away from them), and become more expensive to access.

Another interesting fact about Canada is the physical size of the country and the horizontal arrangement of the population. Of almost 10,000,000 km^2 (9,980,000), about 9% (890,000) is water. Also, the majority of the population lives within 200 km of the U.S. border, with the country stretching to a width of 6,500 kilometers. Wide geographic distribution across areas with very diverse physical terrain and resources has, over time, created many regional cultures and economies within Canada. There may be as big a difference between provinces in Canada as there might be between some European countries. This variety is naturally reflected in the development of the use of horses, and the different horse activities and cultures that we have today.

The unsustainability of such large and rapidly growing cities is only beginning to be noticed in Canada. The level of concern for supporting endogenous rural development, and equal access to resources needed for well-being such as Health Care, education and job opportunities, such as we see in countries like Sweden, is not present to the same degree. In fact, there are trends toward removing services from areas with low population on grounds of insufficient economies of scale. For horse facility owners who depend upon a local population to buy riding instruction, horses or hay, the further decline in population which occurs when standard government services are reduced, has a negative impact on the ability for the horse business to continue in that area. It is a downward spiral or negative feedback loop noticeable in provinces with lower population density and smaller economies. For example, such weakening of rural economies through population decline and service reduction is happening quite noticeably in Nova Scotia.

However, there is recognition that with an average of only 6% of newcomers to Canada choosing to settle in rural areas (Statistics Canada, 2011), there is need to attract people to rural areas (OMRA, 2014). A number like 6% indicates and underlines the predominance of Caucasian access to horses. As immigration policy and ministries of rural affairs work to bring new groups to the rural areas, the face of horse-ownership and equestrian practice will experience even more change.

Regional trends

Depending on the province, horse owners in some areas tend not to live with their horses, but to board or keep them in a livery. This is the case in Ontario (Evans, 2011). In areas where rural economies are more robust, and there are higher levels of the population who are entrepreneurs, horse owners tend more to be able to live with or near their horses. An example is Alberta. The ability to keep horses on your own property or to access them more easily or for more of the population to have exposure to horses, results in a higher place of the horse in the general regional culture such as is the case in Alberta. The opposite, where nearly 50% of horse owners must commute to their horse from an urban area such as in Southern Ontario (Evans, 2011), corresponds to a very weak popular affinity for horse activity in Ontario. Such correlations are not documented academically in Canada. There is a wide range of topics which are neither formally studied, nor formally documented. There is only one University in Canada which conducts equine related research, most of which is in veterinary or animal behavior areas.

In such areas, horses are not part of a central economic activity, such as beef farming, so they are more peripheral as a consumer leisure option. In Central Canada, horse events are very rarely visible in the media or attended by the non-equestrian population (as spectators). Horse activity is practiced by those who already are familiar with it, and there are not many opportunities for the general population to experience horses or appreciate that there is even a sizable horse industry in the area. Popularly accessible activities such as renting a horse to go on a trail ride has become almost non-existent in Ontario, whereas forest/mountain trekking with horses and dude-ranch visits are still more common in Alberta.

Different trends in ownership patterns and land availability have an impact on the type of activity chosen. For example, recreational trail riding and backyard horse ownership are very popular in Western provinces. In Central Canada (primarily Ontario), traffic levels, population levels, land expense, and other factors result in more tendency for practice of equestrian activity to be limited to those sports which can occur in relatively small space, such as the English and Western competitive disciplines. This type of trend creates a dynamic and culture of its own. Where there is more emphasis on equestrian practice for competition, there are corresponding economic and cultural drivers shaping the evolution of the industry and who has access to it in that area. Where equestrian activity is not such a visible part of the broader culture, it becomes quite easy for those responsible for governance and planning to forget that equestrianism needs to be factored into their development plans.

Horse farm location and urban-rural challenges

In peri-urban and rural areas especially near large cities, the place of the horse in recreational spaces, for example the few recreational nature trails available, is in dispute. There are many competing agendas for these spaces, including motorized vehicles such as ATV's or snowmobiles, dirt bikes, walkers and hikers, and others who will imagine that they should not have to deal with the possibility of their enjoyment interrupted or ruined by the presence, or unsafe behavior of a large animal perceived to be the pet of the wealthy who can afford their own land. There are disputes by user groups in most areas over whether to share trails or have separate trails.

Also, even without issues of perception and competition for nature resources, horse owners face risks in using their horses off of private property due to lack of public awareness in how to drive or conduct oneself around animals. Except for areas with strong presence of Mennonites, Amish or Quakers, rural roads are not planned with a horse or bike lane and it is quite dangerous to be on them, except if you are in a car. When horses or cyclists are encountered by motorists, car drivers drive too quickly, not understanding they can spook a horse or a dog. They also drive too close to anything on the road smaller than they are. Road signs notwithstanding, the general population does not expect and is not prepared to adjust their driving to the presence of other road users. This problem has worsened over the years as the increasingly urbanizing population has less and less experience with driving in the country, or with animals. As you can imagine, this would have a large impact on riding in heavily built areas, as well as an impact on carriage driving.

Carriage driving in Canada other than in enclosed areas has declined to almost non-existence outside of Amish and similar communities, and those few horses which are used to taxi tourists in designated areas of city centers. Naturally, sleigh driving and skijoring are affected similarly and can almost only be practiced in privately owned lands, where there is a lot of snow. These winter driving activities are not competitive activities in Canada, and it is next to impossible to know how many people and horses are involved or if anyone practices them. The land used in Canada and in the United States is largely either privately owned, or wilderness space. The concept of the Commons which exists in the UK, or of public right of access to rural resource such as exists in Finland – for example to walk or collect mushrooms – is not in place in Canada. This greatly reduces the opportunities for horse owners to utilize rural spaces, and for non-horse owners to enjoy rural experience and perhaps experience horses by walking through or past their fields, or encountering them on the same trails.

The physical, economic, and conceptual distance between those who practice or are involved in equine industry, and those who are not, is significantly reflected in the list of the top 10 most popular sports in Canada. In a Statistics Canada survey in 2010, the top 10 sports practiced by Canadian adults were: golf, ice hockey, soccer or football, baseball, volleyball, basketball, downhill skiing, cycling, swimming, and badminton. This is very interesting, especially when you consider that over 65% of equestrians in Canada are middle aged adults. It is clearly an activity of choice for a middle aged, female, middle class sport consumer and generates billions of dollars (Evans, 2011), yet still does not list in the top 10 sports. I'm very surprised that even badminton – I do not know any-one who plays badminton other than in their own backyard or while camping with family – even badminton scores higher than equestrian pursuits.

Equestrianism and the Canadian leisure landscape

According to the Solutions Research Group in Canada (2014), there has been a shift between 2010 and 2014 in which sports were popular. In their study on youth participants, sport popularity in 2014 was in descending order: swimming, soccer, dents, hockey, skating, basketball, gymnastics, track and field, ballet, and karate (www.srgnet.com). This is a very interesting selection because it shows the transition from a majority of ball sports, to greater practice of a wider variety of activities in which skills are developed for general physical literacy skills which are also highly transferrable to riding. For example, the balance and proprioception developed in dance, skating and gymnastics are highly transferrable to riding. This list also shows a higher percentage of activities which are enjoyed by a majority of female youth participants. However, equestrian is still not in the list.

To compare numbers, the number of youth in the 3-17yr age-group participating in hockey, skating and gymnastics is just over 530,000, 435,000, and 335,000 respectively. Rate of participation in equestrian or equine activity is estimated at around 400,000, and again with less than 100,000 being youth in that same age category. Another interesting comparison for understanding the landscape of equine activity in Canada is a comparison between the number of coaches (approximately 1,500). In the popular sport of skating there are approximately 5,200 coaches, all of whom would be trained, certified, and registered with the national body with requirements of professional updating. Canada does very respectably in international competition with skating.

Industry growth drivers and developing equestrians

According to figures I had access to while working with the national equestrian federation, we estimated that there were upwards of 5,300 equestrian coaches in Canada. However, when I inquired in preparation for this report, less than 1,500 coaches were listed with the national federation as certified and active. This means that while education and skill development and opportunities for participation in other sports in Canada are highly organized, the situation is not the same in equestrian activity. When you compare this situation to countries where a much higher percentage of horse instructors have formal training for introducing people to activity with horses and for building athletic skill, you begin to see some of the challenges that the equine industry in Canada faces in providing consistent quality of experience and training for the future growth of the sport both at elite and recreational levels. Consumers in Canada have very high expectations for the professionalism of instructors in their recreational activity choices, especially for their children.

As can be seen, the dispersal and distribution of locus of control of the industry is a theme which affects all parts of the industry. As previously mentioned, the national federation activities are directed by over 70 committees. Aspects of the industry which are not related to FEI sport, breed production, and recreational participation are only very loosely associated with the national federation. For example, the racing and tourism industries are not highly connected to the equestrian federations. So, while in other countries revenue from the racing industry is a significant form of investment in other aspects of the industry, this is not the case in Canada.

Interest and impact of racing

In Sweden the racing industry subsidizes lessons schools where children learn how to drive ponies. It is not expected that a high percentage of these children will become professional race drivers. However, the subsidization allows a horse experience for youth which is priced comparably to other activities such as gymnastics or dance. This makes the activity accessible to families from a wider socioeconomic range. Also, each family which brings children to the school is then exposed to the sport of racing. This in turn promotes popular awareness and enjoyment of racing. The race grounds that I observed outside of Skara had a child playground, turning the place and the event into a family event. Racing in Canada is not a family event.

Secondly, the possibility for benefits to the equine industry as a whole from the racing industry in Canada are greatly reduced by the fact that unlike countries for which racing is the only or the main gambling option, gambling on horse racing competes with other more popular gambling and gaming options in Canada. At one time in recent decades, gaming machines were introduced to racing locations to try and inject money into the racing industry from a separate gambling activity. It was very interesting and very sad from the perspective of the equine appreciator because you could go to a racing location and notice that the people were strongly engaged with the gaming machines but were not watching the horses. While they may have also been betting on the horses, the lack of engagement with the horse activity itself eventually resulted in greatly diminished public interest in horses.

For most urban people over many years, the main opportunity to engage with horses was racing. The era of popular and great films about race horses and their ability to unite a country and inspire the population is gone. Walter Farley, the American author of the Black Stallion series would not be a best seller today. Such books have been surpassed in the general population by fantasy and other genre. Most recently in Canada, the government, which runs gaming and casino locations as well as lotteries, has made the decision to remove the gaming from the racing locations. The racing locations are now viewed by policymakers as a bygone activity and an unnecessary drain on the possible profits from gambling.

The position that horse racing has in Canada is significant for the rest of the industry. Many people involved with recreational and sport horse activity are less conscious of the importance of the racing industry to all who enjoy horses. While this importance has not been made through initiatives such as the sponsorship of youth schools for equine activity, the influence has been more indirect through other industry aspects such as: veterinarian research and advances; increased economies of scale for veterinary and horse care products; and increased economies of scale for nearly every other equine associated good such as hay, feed, equipment and even trucks, trailers and tractors. Another effect of decline in the racing industry is a growing proportion of waste horses to the meat industry.

In terms of numbers, the racing industry represents $5.7 billion, 70,000,000 in provincial and service tax revenue, 45,000 horses (58% of which are in Ontario), 27,310 races (68% of which are in Ontario), $321,000,103 in purses paid out (86% in Ontario), 1.45 billion wagered (1.04 billion in Ontario) (Evans, 2012). Racing accounts for 62% of Horse-related jobs nationally, and just over 5% of horse sales. It is clear that there is a very high disproportion between jobs generated by the racing industry and horses involved, especially as compared to pleasure and sport use. It is also therefore easy to see the kind of economic impact that decline in the racing industry would have for horse related jobs.

Horse use distribution

In Canada, there are very few who raise horses specifically for meat. Most meat horses come from other aspects of the industry from which they have been rejected for some reason. In regions with high population density like Ontario, most meat horses are unwanted horses: horses from the racing industry, or which for example have been kept in someone's backyard who had very little knowledge of how to keep the horse and consequently the horse developed behavioral issues. Also of course there would be many horses entering the slaughter stream due to lack of further utility as a school horse, or the owner's inability to pay for vet care or euthanasia.

It should be noted that what has been said about the racing industry and lack of communication between different aspects of the equine industry as a whole is a generalization. There are certainly regional differences in this regard. For example, the British Columbia Horse Council has done a lot of work bringing different stakeholders together to work together for advocating in the best interests of issues affecting all stakeholders, such as land zoning, feed crop supply, legislation and policy, and presence of horses in the mind and experience of the population at large (Horse Council British Columbia, 2010). Perhaps as a result, the equine industry in British Columbia has a much stronger position in popular awareness, and a much higher participation rate in relation to the general population than can be said for other provinces. There are no bodies, academic or otherwise who study such questions, but it would be very interesting if more information and analysis were available.

Of course it is important to share some of the numbers of the industry as a whole. The equine industry in Canada is a $19.6 billion industry which represents 154,000 industry and related jobs, of which 76,000 are on-farm jobs. There are also approximately 963,500 horses, of which 77% (or 744,000), are mature and in use in sport recreation and breeding activities. Compared to many countries, this is an enormous wealth in horses.

As in many other places, the horse industry is heavily dependent on non-commodified subsidy in the form of unpaid for volunteer time. There is an estimated total of 400,000 fulltime equivalent positions which are filled by volunteers. Herd size in Canada has remained relatively stable in the past 15 years with a slight rise around 2005. However, it is expected to drop by 30% over the next five years based on surveys of owners and their intentions regarding horse ownership (Evans, 2011). The main reasons cited for expected drop in horse ownership are: expense of keeping the horses, age of owners and horses and difficulty of finding buyers. The equine industry population is aging out, as mentioned.

Horse farming: falling between agriculture, hobby and rural recreation enterprise

In Canada, horse farming is not generally considered agriculture. This means that it is not grouped with other agricultural activities in policy or other administrative considerations. In the government Census, farms are counted only if there is a minimum of $5,000 in a year in livestock sales. In Canada, this revenue criteria encourages breeding of horses for sale, quite apart from sport and recreation needs or quality consideration. In order to qualify as an agriculture business, many horse farms find other creative solutions. Some examples would be raising of other livestock species, or hay or crop sales. An interesting fact is that horses are the only livestock which is subject to a goods and services tax normally applied to products and services. Government revenue from this tax totals approximately $145,000,000 – a revenue source which is not likely to be given up any time soon.

Because of the definition of a census farm, Canadian government reports are not a reliable source for accurate data on the size of the industry. For example census data from 2006 showed 54,169 farms. In Evans' report of 2011 (Evans, 2011), it was estimated that there are more than 145,000 horse farms. Similarly, the 2006 census showed 453,965 horses, whereas the Evans study estimated 963,005. Also, conservative estimates of participant numbers are around 400,000 (undisclosed private sources of information). However, the Evans study estimates a number upwards of 850,000 (Evans, 2011). This discrepancy is present across the whole equine field, from populations to economic impact. More research and documentation which can deliver concrete numbers is extremely vital to the sector and for public policy development. Other information in the Evans report indicates that there is an expectation of a significant drop in equine industry participants over the next five years. This point will be explored a little further later.

In Canada, the average size of a horse farm is 275 acres. This average is, however, derived from a very wide range. For example the average farm size in some areas in the West is 900 acres, which would obviously suggest that the farm was also engaged in cropping activities. At the other end of the spectrum, farms in Newfoundland average 10 acres. This makes sense because Newfoundland has very little arable land. In the province of Ontario, the average farm size is 80 acres. Differences in farm size depending on region will certainly have an impact on the style of horse-keeping which is adopted.

Horse farming costs and revenues

Other costs include the cost to feed a horse per month, which ranges across the country between one and 200 Canadian dollars per head. Obviously it is less expensive to feed a horse on pasture where that is available, than to feed a horse in an area of high population density, with little pasture and greater dependence on processed feeds and small hay bales. In Canada the most used forms of hay are small square bales, and round bales for outside feeding. Hay cubes are also used. However, haylage such as we see in Sweden and Finland is not common. One reason may be that in spite of the seasonal stress of the farmers and horse owners, there is actually a fairly reliable season for cutting hay and allowing it to dry in the field.

Horse boarding or livery costs are also obviously impacted by the geography and predominant aspects of the regional horse culture. Boarding fees can range between $350 to $550 Cdn. per month per horse, with competitive facilities and facilities closer to major urban centers costing more. The total overall estimated care cost for keeping one horse is approximately $2,700 per year outside of boarding fees (Evans, 2011). The average horse purchase price is $6,000, and again there is obviously a wide range. For example in Ontario the average is $9,000, whereas the average in Manitoba is $2,500. It is possible to buy a horse at the meat auctions for $350 to $500. Horses rejected from the racing industry are also often very inexpensive, in the range of $1,200.

With such a range of horse prices and such a cheap meat price, it is easy to see how certain issues arise. For example, there is a problem in New Brunswick with welfare of backyard horses. The province is very rural and it is still part of the culture to have animals in the backyard, even though these animals may not be used very actively in recreation and certainly not in competitive sport. When meat prices are high, it is more attractive for people who do not want or need their horse to sell the horse for meat. When the prices are very low, the owner would weigh the trouble and costs of transporting the horse to auction against the gains. The result in a lower economy is a larger number of horses which are not being used, and are also owned by people that are not putting a lot of money into the horse. It is not difficult to see that if a person calculates they cannot afford the cost for trouble of transporting the horse for meat sale, that they do not likely have a large budget for veterinary or Farrier fees. This is a problem in some areas. The results can have an important impact on horse welfare.

At the other end of the spectrum, a very wide range in horse sale prices is indicative of significant dependence of the buyer on the marketing ability of the seller. It must be remembered that Canada, like the rest of North America, is a marketing and consumer goods driven economic culture. In that context, it is important to know that unlike in European countries, Canadian horses do not all have passports. In fact a very small percentage of Canadian horses which are used in FEI discipline competition have passports. Race-horses would be an exception however, because of the requirements for recordkeeping which are quite different than for pleasure, breeding or sport horses.

Records and management

Outside of racing, breeding associations keep their own pedigree records. However unlike a European horse passport, this pedigree record does not always include information about the horse's offspring or competition results, nor the competition results of the horse's relatives and offspring. Such information is certainly not available publicly. A breeder would keep private records of competition results. A buyer would be entitled to only the information which a breeder chose to share, or which might appear on the pedigree. Combine this data storage pattern with a habit of using horses not active in competition for breeding, and it is possible to see the impact on quality. A common practice for more amateur horse breeders is to take a mare that has not done well in jumping, and market her as a dressage horse, or breed for dressage. Horses which have foot and other issues preventing a robust riding or competition career are also sources of breeding stock. The results for the talent pool can be very negative.

Of course, there are many serious breeders, and many good horses. For example, the breeders of Arabians and Quarterhorses have robust breeding programs, with the Arabians in particular having pedigrees that go back hundreds of years, and these groups have been selling good horses internationally for some time. Thirty years ago, the thoroughbred was still the most popular English saddle discipline sport horse in Canada. Now, warmbloods are becoming more dominant and warmblood breeding remains a fairly recent activity. Because warmbloods are essentially mixed breeds and the activity is young, many breed associations do follow European requirements for breeding stock approvals prior to breeding, and keep good records, even if there is no national requirement to match sport performance with breeding stock.

However, there is no restriction or control on what a buyer of these horses does with them. A Canadian buyer can buy an Oldenburg or Hannoverian mare that didn't make an inspection score, breed her to whatever she wishes, and sell the foal to whatever buyer she attracts. It might not be purchased by a serious buyer, but it will be in the gene and competition pool. There is also some amateur breeding of cold-blooded horses to Thoroughbreds to produce what are then referred to as warmbloods. There is a breed association governing quality of these Canadian warmblood crosses for registration to some degree, but there is no limit on what can be cross-bred and marketed to a willing buyer.

Very few breeders take the time to also track the competition results of the offspring. Horses which are not pure-bred, or at least professionally bred crosses such as warmbloods, may not even have a pedigree. This situation leads to some very interesting practices which would be different from what happens in other countries. For example in the hunter discipline, it is quite common for horses to be given a show name. The show name might change if the horse changed owners – the show name is necessary because the horse does not have a pedigree in the first place and has not been given a name at birth which is registered somewhere.

Breed associations are not required to report horse numbers to the national federation, so there is no single organization which tracks overall herd size or composition. There have been national reports commissioned twice, including the one referred to often here. When such research is done, only approximate numbers are generated. From the perspective of the way that horse pedigrees and passports are organized in Europe, it is possible to see how this diversified organization of horses easily results in the lack of ability to be precise about total herd numbers, breed numbers, or equine athlete talent pool. It is also quite impossible to control quality and elements such as congenital factors which are not desirable for horses. Buyers are very much at the mercy of the marketing abilities of the sellers.

This is obviously a generalized comment. There are clearly many very knowledgeable breeders, owners and trainers in the equine industry in Canada, as well as many very good horses. Canada has also recently produced a celebrated code of practice for animal care which includes a very complete section on horse care guidelines (National Farm Animal Care Council, 2013). It remains to be seen how much this influences practice as it exists now.

Governance, horse care and horse meat

Anyone in Canada may obtain a horse or two and put them in their backyard property, provided that the property meets the loose requirements of size of land needed for two horses, and zoning requirements. Then the person may breed the horses regardless of how much knowledge they have about horse care, breeding and training. The obvious result is a very robust meat industry in spite of a very low occurrence of official horse meat producers. Canada is the third largest producer of horse meat in the world, importing 63,688 head, and exporting 21,090 head and 17,700,000 kg of meat worth $90,000,000 (Evans, 2011). The top five markets are Switzerland, Japan, France, Belgium and Kazakhstan. Horse meat is not consumed in Canada except for in Quebec, and is not tracked in the Agriculture Canada red meat figures which do include deer and elk which run a lower volume since they are smaller animals and only 52,051 head/year at the last count (Agriculture Canada, 2014; Canadian Meat Council, 2013).

Horse meat is a public hot-potato issue, as well as an animal welfare concern. Given the source of the horses and their lack of passports, it is not a meat which can be relied on to be free of medications and hormones to the same degree as other major products such as beef. The requirements on all food for human consumption are very strict in Canada, but there are no health issues occurring with horse meat. However, from a welfare perspective, the inspectors of processing plants are the same as for other, larger meat products, and they are quite busy. There are issues in Canada with groups protesting the welfare of the horses in transport and slaughter, but it is not an issue which people like to discuss openly. As soon as it gets in the media, it disappears again and is forgotten quickly. Horses transported from the US, which has no processing plants, sometimes travel very long distances.

While not necessarily a popular position, the official position of the national federation is that horse slaughter is at least one viable end-of-life option for horses. In acknowledgement that not everyone can euthanize horses they do not wish to keep, horse slaughter is viewed as a better macro-industry option than the option of having tens of thousands of unwanted and uncared-for horses. In Canada there is not the 'unwanted horse problem' one hears talked about frequently in the US. Equine Canada is the one body which advocates to the government and the Canadian Food Inspection Agency for improvements in execution of processing control policies.

Tracking and usage of horses

To come back to the general industry, the division of uses of horses is as follows: working horses 5%, meat 5.5%, breeding 5.9%, racing 7.3%, sport 35.1%, and pleasure 41.1% (Evans, 2011). The division of sport and pleasure is not precise. This is because this information is based on surveys dependent on participant understanding of what they are doing with their horses. Sport license numbers with Equine Canada seem to indicate that a much smaller percentage of horses are engaged in sport. However, Equine Canada does not govern most of the Western discipline competition activity. Also, respondents might consider themselves as active in sport who are actually attending a notational local fair or competition. In addition, provincial equestrian federations contacted indicated estimations of leisure activity around 85%. Roughly 80% of people selling horses make private sales. Private sales may not be their exclusive channel, however it is still the dominant vehicle for horse sales.

Once the horses are sold it is interesting to know the purpose for which they are obtained. Again, it is difficult to be precise and the numbers come from survey data collected from the Evans report. The breakdown is as follows: sale of young horses at 22.8%, sale of breeding stock at 18.3%, sport use other than racing at 18.2%, and pleasure riding or driving at 17.8% (Evans, 2008). Only 2.3% were indicated as purchased for riding lessons. This is a very significantly low number, which seems to correspond to anecdotal reporting of drops in numbers of riding schools in recent decades. Many facilities do not find it cost effective to have school horses. In Canada riding schools are mostly privately owned horse facilities, and have no subsidy from municipal or other gov't as a recognized municipal sport activity. Horse ownership and care costs must be covered on purely private enterprise business models. An increasing number of riding schools prefer to have mostly private boarding clients who take lessons, or at most lesson horses which are also in part-lease. It is less and less common to find a riding school with enough school horses to run several lessons in a day.

One reason for decline in presence of riding schools can be related to the care costs previously mentioned, in contrast to a possibly declining market. The aging demographic in Canada is very significant. In 2003 the ratio of child participants to adult participants in equestrian sport was 69 to 31. By 2010, the ratio was 41 to 59. More recent estimates show adult participation at well over 60%. The market has been shifting from servicing group lessons for children who quickly outgrow ponies and horses, to servicing the middle aged, middle class female client with disposable income for lessons and horse expenses. Current demographics of equestrians in Canada showed a median age of 50 to 59 years old, 84% post-secondary educated, a median house-hold income of $60-80,000, and a predominance of female participants at 79% (Evans, 2011). The general environment is more of a private ownership and consumer choice option than a club based or Community-based activity.

Globally, practice of equine activity is dominated by the English saddle disciplines at 54%, Western saddle disciplines at 23% and the remaining 23%, bringing in everything else. Less than 1% of participants practice polo, saddleseat or vaulting. Reining and competitive trail or endurance account for 2% each. Within the English saddle disciplines, hunter/jumper is very popular at 27%, dressage follows closely at 20%, and eventing accounts for 7% (Evans, 2011). These numbers are based on self-identification which means that they are of limited precision.

In terms of the distribution of competitive equestrians, the numbers are calculated from Equine Canada sport licenses by type. In some cases the proportion of participants is reflective of the Canadian population distribution. In other cases it is due to the degree or lack of popularity of horse activity in the general culture, and the relative economic health of the province. To the numbers are as follows: Ontario 41%, Alberta 17%, British Columbia 17%, Quebec 14%, Manitoba 3%, Nova Scotia 2%, Saskatchewan 2%, New Brunswick 2%, Prince Edward island 1%, the Yukon Territory less than 1%, Northwest Territories less than 1%.

In terms of the type of activities which are practiced in each province, these are of course influenced by culture, economy and other factors such as prominent individuals who invest in a sport, or significant facilities which attract a hub of activity. For example, there are excellent and well known jumping competition facilities in British Columbia and in Alberta, such as Spruce Meadows. Consequently, hubs of activity in jumping have arisen around these areas. A summary of the key activities, or activities that are unique, by province is provided in Table 1.

Tracking competitive horses

The equine Canada sport license system grants eight sport license appropriate to the level of competition of the holder (human). Thus a rider would purchase a sport license appropriate to the horse they are competing with in the highest category regardless of how many other horses they compete with in other categories. Therefore sport license levels are indicative of the human competition level and not the horse population. People in beginner categories might not purchase a sport license if they are only attending local events which are not sanctioned or governed by Equine Canada.

Table 1. Regional distribution of equine activity by Province. Colored cells indicate that activity is significant in a province.

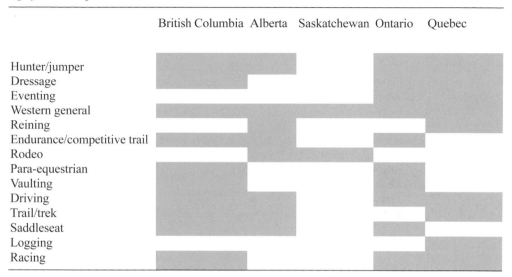

	British Columbia	Alberta	Saskatchewan	Ontario	Quebec
Hunter/jumper					
Dressage					
Eventing					
Western general					
Reining					
Endurance/competitive trail					
Rodeo					
Para-equestrian					
Vaulting					
Driving					
Trail/trek					
Saddleseat					
Logging					
Racing					

The competition levels are theoretically organized by level of difficulty. However, in practice they can often be more descriptive of the size and importance of the competition, or the competitor's preference for location, rather than the actual technical level of riding. For example it is possible to compete in hunter over 2'6' fences at a local or bronze show, as well as a provincial level or silver show, or a so called national or gold show. The same is true for dressage where a rider may compete with a horse at Level 1 (the American system is followed instead of at the lower levels FEI) in a national, provincial or regional horse-show. The federation is currently reviewing this situation to integrate the long term athlete development model into the competition system.

In Eventing and Endurance there are rules governing progression of competition and allowance of horses up the levels based on technical performance. With an understanding of this system of competition organization which is largely influenced by geographical proximity of the competitor to the show grounds, the distribution of sport licenses cannot be interpreted as an accurate description of distribution of technical skill. Sport license distribution is at 43% for a gold license, 20% for a silver license, 28% for a bronze license and 6% for a platinum license. Since no one would have a platinum license unless they were competing at the FEI level, platinum license numbers are a more likely fair representation of ability. Participation in competition is more influenced by financial consideration.

The disproportionate number of participants at gold competitions can be partly explained by the need for coaches to fill their horse vans with clients as a cost effective way to cover costs of their own competition at these shows. With classes of many levels at the show, the coach can compete in their elite level, while the clients compete in lower levels. Clients with money prefer such shows because the amenities are usually nicer and it is more prestigious and a better business model to say that their child needs a very expensive pony because she is competing at national competition. In reality, there is also a higher likelihood of higher quality horses to compete with for the client. The small talent pool of elite riders and horses in Canada means that these riders have very few events in Canada featuring the type of rigorous competition which would prepare them for international competition. In consequence, most of them compete at least part of the year in the United States or Europe.

To return to the subject of riding coaches, it has already been pointed out that a large percentage have not been formally trained in coaching skill, or if they have, do not adhere to the voluntary system of ongoing professional updating up and accountability. Just about anyone can claim to be a riding instructor or trainer, create a business and operate as long as they can find clients. There are certainly many highly trained coaches who have learned from masters elsewhere in the world, or who have gone through formal certification and training programs in Canada to be competent at teaching the majority market which is an adult amateur rider. Interestingly, the Evans study showed 55,200 professional horse trainers in Canada. This is interesting because there are only just over 5,000 coaches in the historical Equine Canada database, and 300 instructors certified through the Certified Horsemanship Association, Canadian branch. When I worked with equine Canada we conservatively estimated that for every record and certified coach there was at least one instructor practicing without certification. This discrepancy points once again to the lack of precision in the knowledge of the industry, and challenges with its direction.

Rider participation

In the Evans study (2011) it is estimated that there are approximately 43 students per instructor. However, informal polling of coaches shows an estimate of around 100. Most coaches are not fulltime professionals, and have an average coaching revenue of $6,000 per year. Of the coaches who are certified through equine Canada, 87% would be instructor or level one equivalent which means that the right ratio of coach-to-student is in place to meet the predominantly amateur market. Since riding facilities are privately owned, there are very few jobs for fulltime coaches. Equestrian coaching is not delivered by any university in Canada. Most coaches are female, and most fulltime coaches practice out of the equestrian property which they own or rent. Coaches are not usually full time coaches unless they are also an equestrian business owner. This means that success as a professional coach is very tied to business skill ability.

The very wide parameters of the equestrian industry also allow a lot of room for creativity and entrepreneurship. There is no one model for the type of business skill which coaches must have in order to become successful enough to work full time with horses. And the sector shows a disinclination for continuing to develop their own athletic skill. The pathways chosen reflect the individual. Some may be very good at running a barn and facility and managing clients. Others may be very good at coaching and producing new riders, and some may be very good at networking and marketing to attract sponsors and buyers. This range of business skills is not taught formally.

Development of future such coaches happens when a promising rider is lucky enough to be able to apprentice with a coach or rider who is successful in these ways. Since riding is not a community-based sport, it is usual only for riders with family money to afford lessons from such prominent coaches, and to privately purchase trained schoolmasters, to become their apprentice. There are always exceptions. However, this example simply shows that there is not really a cohesive system for developing coaches, athletes and horses. For the most part, this is not a large problem because the majority of equestrian participants are adult amateurs who mostly participate for pleasure rather than serious competition.

Trends and industry future

Given the shift over the past 3 to 4 decades from the younger market to an older riding market, the future of equestrian sport and horse ownership and the industry as a whole is threatened by the problem of attracting new young participants. Also, in Canada there is a public health problem of obesity and sedentariness, which is experienced by all sports as a problem for athlete development, as well as for consumer choice of active living options. For equine activity, the traditional market is middle aged adults with money and time to be able to ride, or the children and grandchildren of people who rode, ride, or always wanted to ride. This second group is in decline. It is a problem in the industry that marketing of equine product and experience is traditionally aimed at people that are interested or already involved in horses. There is a very low level of experience with marketing outside the already initiated group. As a result, there is a large decline in youth participation which has already been noted, without a recovery plan in place for recruiting new participants who have never before been exposed to horses.

The lack of youth participants is reflected in a drop in pony club membership of around 50% in the past 30 years. Also, provincial equestrian federations have seen a drop in the numbers of members who are under the age of 14 years old. Equestrianism is considered an early start late development sport which means that certain skills are best learned in early physical and mental development brackets. The lack of riders in the six year to 14 year old age-group means that more riders start after the optimal age to develop basic skills of awareness, balance and physical control. In addition, lack of input of new participants into the system means that there is little to replace the people who will leave as they age out. Current marketing trends of focusing on the middle aged female suggest that the industry could be facing declining participation numbers in the future.

Other factors contribute to the drop in participation in riding. As previously mentioned, with the growing distance between the location where horses are kept and the activity consumer, the cost of fuel and time commitment to the activity for a family increases. The Pony Club surveyed members and found that reasons for drop in membership were in order of decreasing importance: cost of fuel and distance, time, other family recreational commitments, finances and time related to break-down of the family unit, less access to clubs which had available school horses, and burn out of the parent who was also heavily involved in volunteering in the pony club.

Approaches to solving the problems of getting the word out to the general population and promoting horse activity are different from region to region. The provincial equestrian federations are more engaged with this task. The province of Quebec is an exceptional example in having quite a robust level of youth participants. Some possible reasons can be related to enhanced support to sports in Quebec by the provincial government – programs such as sport-études which allow young people to ride in the day as part of their study program just as their peers might do in hockey; and the fact that the Quebec federation has a very strong regional club structure and grass-roots competition system. The regional club and grass-roots competitions result in local events and a community cohesion, or a richer atmosphere of being part of a group having fun.

The Quebec popular culture is also somewhat enamored with the American cowboy image, an image which is considerably more `cool' than English breeches and boots, as well as more accessible to a middle-class or working class public. Trail riding and leisure riding are more culturally common as well. Quebec and Alberta are the two main hubs in Canada for Western riding, and specifically for Reining. The Quebec federation also developed the only curriculum for very young riders (a picture book with tests, levels and interactive web-page). It has since been adopted in many other Provinces.

Conclusion

There are notable gaps in this report in the information provided, as well as the range of references used when compared to the type of background research and literature body which would be available in some other countries. This is because there is no real body of literature on the industry, apart from the Evans reports. The Evans reports are the product of a private consulting commission, without peer review and without triangulation with other documents. The various Evans reports are the main body of data available on the overview of the industry. The national Federation, Equine Canada, does not sponsor or conduct academic research, has very little control over data collection by the diverse contributing bodies, and does not systemically track information usually tracked by other world equestrian federations. Other information gathered was largely from government or other organizations, some of whom do not publish their information, or do not collect data on various pertinent questions. Ironically, the business adage 'you can't manage what you don't measure', does not seem to have caught on strongly, at least not in terms of meta-trends.

Preparation for this short presentation was very challenging and time intensive, involving a great deal of personal contact with organizations and use of their unpublished and sometimes undocumented, observations. The relative absence of robust academic or industry measurement and study in Canada is in itself a significant indicator of the state of the industry. Driven largely by free market forces, it is hard to say how many of the various problems can be addressed except by engagement and partnership of stakeholders across the country.

While considerable space has been allocated to describing the challenges to the equine industry in Canada, there are of course also many strengths and opportunities. The entrepreneurial and independent nature of the industry can also be viewed as a potential future strength. What may be lacking in systematic approach may leave room for creative adaptation to changing cultural and economic conditions. A lack of antiquated and overly bureaucratic structure can be seen as leaving room for solutions which are based on comparative best practice for more current problems. Also absence of a single and cohesive dominant approach in certain areas needs space for multi stakeholder collaboration. Having regional differences is a wonderful opportunity to observe whether practices work well or not, and under what conditions. Some initiatives which hold promise for positive growth include:
- The adoption in Canada of a long term athlete development model which has been integrated now into the goals for equestrian coaching, rider development and competition system.
- The upcoming Pan-American and World Equestrian Games which one hopes would bring horse activity back into wider public appreciation.
- The fact that equestrian activity can be widely accessible. People with disability can participate, it is one of few sports that attracts a predominance of girls, and offers unique wellbeing benefits through interaction with an animal.

Canada has been a nation dependent on the creativity and innovation made possible by newcomers to this country bringing new ideas and different ways of doing things. The relative high rate of immigration may be changing the demographic picture. However, if the industry succeeds in engaging the current culture and opening its doors to wider participation there is ample room for growth.

Acknowledgements

I wish to thank the organizers of the EAAP Conference in Copenhagen in August 2014 and the Canadian Society for Animal Science for inviting me to present on the state of the equine industry in Canada. This chapter is based on that presentation. It is a great privilege to have the opportunity to share information about the Canadian equine industry in this forum. Sometimes it is tempting to think that the situation for the equine industry in Canada is highly unique. However in having the opportunity to share information, it can be surprising how many issues and trends are actually shared in common between countries. While I cannot comment on the industry in other countries, I can leave it to you to discover commonalities and differences with your own situation.

My own involvement with the equine industry in Canada includes experience as a national program and projects director with the national equestrian federation, Equine Canada, where I was responsible for development of the new coaching and rider education programs and materials. In this role, I had an excellent opportunity to meet with many organizational stakeholders, and to get something of a watertower view of the industry. Preparation for this piece included personal interviews, some access to documents, and recollections of information I had access while working with the federation.

References

Agriculture Canada, 2014. Available at: http://tinyurl.com/ne3a3kx.

Brodhagen, H., 2009. Through the eyes of a child: first nation children's environmental health. Union of Ontario Indians Anishinabek Health Secretariat. Nipissing, Ontario, Canada. Available at: http://tinyurl.com/pfu7e95.

Burns, A., D. Bruce and A. Marlin, 2009. Rural poverty discussion paper. Rural Secretariat Agriculture and Agri-Food Canada, Government of Canada. Available at: http://tinyurl.com/qbbq22r.

Canadian Meat Council, 2013. Horse meat production Canada. Canadian Meat Council, Ottawa, ON, Canada. Available at: http://tinyurl.com/n9gg45x.

Cloke, P., T. Marsden and P. Mooney (eds.), 2006. The handbook of rural studies. Sage Publications, London, United Kingdom.

Evans, V., 2008. 2007 Ontario racing and breeding industry profile study. Strategic Equine, Newmarket, ON, Canada. Available at: http://tinyurl.com/pxltpx5.

Evans, V., 2011. 2010 Canadian equine industry profile study: the state of the industry v2. Equine Canada, Ottawa, ON, Canada. Available at: http://tinyurl.com/qyoy56a.

Evans, V., 2012. The economics of horse racing in Canada 2010. Equine Canada, Ottawa, ON. Available at: http://tinyurl.com/q7lcne2.

Foot, D.K. and D. Stoffman, 1996. Boom, bust and echo: how to profit from the coming demographic shift. Macfarlane Walter & Ross, Toronto, ON, Canada.

Horse Council British Columbia, 2010. Equine industry study. Horse Council British Columbia, Aldergrove, BC, Canada. Available at: http://tinyurl.com/qy2pvwk.

National Farm Animal Care Council, 2013. Code of practice for the care and handling of equines. Equine Canada, Ottawa, ON, Canada. Available at: https://www.nfacc.ca/codes-of-practice.

Ontario Ministry of Rural Affairs (OMRA), 2014. Rural roadmap: the path forward for Ontario. Available at: http://tinyurl.com/o2vf3ak.

Solutions Research Group, 2014. Canadian youth sports report. Available at: http://tinyurl.com/l67p2sp.

Statistics Canada, 2011. Population, urban and rural, by province and territory (Canada). Available at: http://tinyurl.com/kfu6ep2.

Part 10.
Conclusion

21. Conclusion to the new equine economy in the 21st century

C. Vial[1,2]* and R. Evans[3]
[1]French Institute for Horse and Riding, IFCE, 19000 Arnac-Pompadour, France
[2]National Institute of Agronomic Research, INRA, UMR 1110 MOISA, 34000 Montpellier, France; vialc@supagro.inra.fr
[3]Norwegian University College for Agriculture and Rural Development (Hogskulen for landbruk og bygdeutvikling – HLB), 4353 Klepp Stasjon, Norway

A new horse industry in the 21st century

In the European Union, less is known about the horse industry than you might think, despite its recent growth and evolution. There is, however, a growing consensus that it has changed from a primarily agricultural and industrial sector activity to one firmly rooted in sports, leisure and consumption. These evolutions generate the development of new kinds of activities and the growth of the whole horse industry.

We can notice an increase in new uses of horses for leisure riding, tourism riding, ethological riding, therapy or social rehabilitation. Practitioners also have new objectives such as the reduction of stress, the creation of emotional relationships with horses, practicing horse riding with comfort and for pleasure and being independent and free in their activity.

The modern horse industry has some particular characteristics, for example many horse owners and breeders are non-professionals and breeding is largely managed as a recreational activity; however low profitability is a significant problem for many equestrian structures all over Europe; and finally, the horse industry is at the intersection of three sectors: the agricultural, touristic and sport/leisure sectors.

The horse industry has various impacts on territories. First, it has an economic significance. For example, estimates are that the equivalent of 400,000 full time jobs is provided by the sector. Second, horses contribute to the maintenance of landscape. They use at least 6 million hectares (ha) of land in Europe (in 2007, due to grazing and food production (European Horse Network, 2007); against 3.5 million ha in 2000 (EU Equus, 2001)). Thus, they are significant actors in the use of rural land across Europe. And they can be used in complement with cattle or sheep as grazing systems which increase the biodiversity of margin and abandoned farmland. Terms of third, horse-related business activities generate an increasing share of income in rural economies all over Europe thanks to the diversification of farmers, the use of horses in modern agriculture and forestry, the production of horse feed, etc. For instance, revenues in agriculture related to the horse industry have been estimated at 4% of the total revenues (EU Equus, 2001). The multiple uses of horses are also important in terms of land occupation and management. Finally, there is the social importance of horses through horse riding, which is a popular leisure activity among children and young adults – particularly female ones, but there is also riding for the disabled, and the creation of new social links around riding and horse ownership and between urban and rural inhabitants.

A wide potential area of research questions

The growth of these new activities generates new questions about the role of equines in economic dynamism, culture, social links and rural development – questions which reflect major changes in society. What follows are some questions generated by the research contained in this book.

First, we wonder how much economic activity horses inspire, and what role they play in the cultural heritage of European regions.

Second, horses are now part of a 'consumer services economy' to provide new services and we need to understand new consumer demands. This also means that horses need to be bred with new characteristics allowing them to become maximally optimal for these new activities.

Third, new uses of horses mean that we must see and interact with them differently, and change how we must view equine welfare. And, in particular, we can ponder the trend to not eat horse meat and ask whether it is not necessarily good for horses.

Finally, the growth of suburban location of riding horses creates new challenges about land management and physical planning and we can wonder how horses can contribute to sustainability.

Horses are clearly important across the full range of diverse nation-states in Europe, yet our knowledge of the actual number of horses, number of riders, the actual economic impact of these activities, the actual use to which horses are put, the actual impact of these activities on the environment, and the actual impact of these activities on horses themselves remains less than perfectly clear.

It is now necessary to address these and other questions in order to support the ever-increasing levels of horse ownership and riding across Europe. This knowledge will be vital to better inform policy creation, from a regional to a trans-national level. And it is vital to support those who wish to dedicate their lives and careers to the pursuit of their passion for horses.

The rise of a new equine economy in the 21st century

Traditionally, equine research has been concentrated on Veterinary Science and Ethology. However there is a growing acknowledgement of the need for the considerations of the socio-economic impacts of the horse industry. It is impossible to understand them without understanding the society within which horses are embedded. Horse welfare, population size, behaviours and potentials depend upon those of the societies in which their owners and riders live. Further, contemporary society is changing as it never has before. There is no single 'society'. Different peoples and different places all constitute unique economies, unique social values and mores, and unique formations of the horse industry. To understand the future of the horse sector we must understand these varied social and economic formations.

In this context, and given the importance of and the challenges faced by the horse industry, the number of socio-economic studies devoted to this sector has recently begun to multiply all over the world. The new equine social scientists undertake research, analysis and the development of new understandings of changes in the economy, in cultural values, and in social organization of contemporary society. Working together with traditional equine sciences, we are creating new interdisciplinary knowledge which helps us understand how we got to where we are now, and where the equine sector might go in the future.

What is this new equine economy? We know the 'old equine economy'. This featured an emphasis on equine sport. equine sport (and in particular, betting) generates large quantities of revenue and a focus on this has dominated research on the economics of horses over the past three decades. Now, with the growth of activities in leisure riding, there is a need to understand the economic impact of activities such as equine tourism, equine assisted therapy and pedagogy, non-rural leisure riding and modern horse ownership, in order to re-evaluate their contributions to the new equine economy in the 21st century.

Illustrating the growth of social sciences interest in the horse industry

As we can see in this book, sociological studies are emerging, analysing sociological evolutions or people attitudes. For example, some researchers have studied the feminization of the horse industry, the evolution of the social status of horses in history and the transformation of the human relationship with horses. Authors also deal with social conflict or co-operation generated by horses, and create actor profiles and behaviours concerning breeders, instructors, buyers and consumers. Thus, these authors have made contributions towards an interdisciplinary understanding of the new equine economy in the 21st century.

We study economic issues across a wide range of focuses: from the specifically econometric such as determining the 'economic multiplier effect' of equine tourism activity, through the impact of regional, national and global economies on horse keeping, to the impacts of the fundamental restructuring of economic activity in developed economies and the growth of, for example, the leisure sector. The growth of the horse industry generates new consumer demand and economic opportunities that are important to study. Each segment of the horse industry has its specificities. For instance, racing and betting generate high incomes, whereas breeding faces some difficulties and needs economic support in several countries. Another example is sports and leisure which is a growing sector involving a lot of amateurs who are organized in different ways, using professional horse livery services or having their horse at home and taking care of them themselves. Equine tourism is growing spontaneously across Europe but has been studied very little and promoted even less. Finally the decline of the hose meat sector generates new issues about horse end of life and well-being but also about the maintenance of heavy horse breeds and of their breeders.

Important key challenges are the counting and localization of equines, as well as the collection of other data such as the number of employees, horse riders, horse owners, horse breeders… and also 'the unwanted horse issue', which has been prevalent in the United States or Ireland since slaughter became illegal.

Some countries have started to underline the role that equines are currently playing in rural development: their multi-purpose uses represent a strong advantage in land occupation and management; they help restructuring the peri-urban countryside; they contribute to agriculture diversification and multifunctional development thanks to horse livery, equine services and agritourism; they help maintain relationships between urban citizens and cultural rural life; but they are also creating conflicts over land use with other users in peri-urban areas.

A growing cadre of researchers focusing on human-horse relations, and new collective endeavours representing this

In this context of a growing interest of social sciences in the horse industry, a working group in socio-economy was created in 2010 within the EAAP Horse Commission. Today, it includes 67 members from 21 countries. This group has been created with several goals in mind. First, we would like to create links between people around the world who are working on the horse industry socio-economy to share ideas, research and experiences. Second, the group is identifying relevant areas for future research and collaboration and building common projects. Third, there is a primary need to improve statistics that measure the size and importance of the horse sector. It will be particularly useful to create links between people in the different countries who are collecting economic data on horses.

When we use the word 'socio-economic' we are focusing in two primary areas – society and economy. Members and participants include economists, sociologists and anthropologists, human geographers, specialists in tourism studies, leisure studies, sports studies, rural studies, ecologists, veterinary scientists and ethologists who realize how difficult it is to treat horses in isolation from the human society in which they are located. Fortunately they, because of their disciplines, many are already used to working in inter- or multi-disciplinary research teams, and this offers hope that together, some of the challenges raised in this book might actually be addressed in time.

This is paralleled by the creation of the Equine Research Network (EqRN) in 2009 at the European Rural Sociology Association in Vasa, Finland. This network now contains over 140 members from over 20 countries all over the world. It is a network of (mostly) social scientists from across the world who study the human horse relationship, who gather together at conferences, share work, and look for opportunities for collaboration.

To conclude, we can say that horses are artefacts and assets of both natural and cultural heritage, that they operate as social links, create important economic opportunities and contribute to diverse and vital rural development. For all of these reasons, the horse industry deserves to be studied. Further, if we join our efforts and knowledge, we begin to see ways to promote its sustainable development. But we need to work together to create true multi-disciplinary research which addresses the complex challenges and opportunities presented by the new equine economy in the 21st century in Europe!

References

EU Equus, 2001. The horse industry in the European Union. Working Report prepared for EU Equus 2001, Skara and Solvalla, Sweden, 45p. Available at: www.horse-web.net/docs/EU_Equus_2011.pdf.

About the editors

Dr. Céline Vial is a researcher in economics of IFCE (French Institute for Horse and Riding). IFCE is a public administered institution which supports the development of equine activities. Céline Vial operates at INRA (National Institute for Agricultural Research). She is localized in France, at the INRA of Montpellier, in the MOISA Unit (Markets, Organizations, Institutions and Operators' Strategies). She also manages the working-group in socio-economy of the horse commission of EAAP (European Federation of Animal Science). Céline Vial holds a Ph.D. in economics about 'An economic analysis of open air leisure and of its territorial implications': the organization of 'amateur' Equidae owners between domestic. Her research focuses on the economics of the horse industry. In the current context of the development of equestrian activities, her work aims at analyzing the organization of equestrian activities and their impact on rural development. The research program on which she has been working since 2006 is named 'horse and territory'.

Dr. Rhys Evans is an Associate Professor of Rural Development at the Norwegian University College of Agriculture and Rural Development (wwwhlb.no). A human geographer, he has been researching and publishing on human horse relations since 2008. His previous publications include chapters on Nordic Native Breed horses, Equine Tourism, horses as economic development resources, and several explorations of the nature of the human horse relationship itself. With Sylvine Chevalier-Pickel, he is the co-editor of the first book on International Horse Tourism, published in 2015. He is a Vice President of the EAAP Horse Commission and co-convenor of their Socio-Economic Working Group. Additionally, he is the co-founder and Convenor of the Equine Research Network – an international network of over 150 researchers who focus upon Human Horse Relations.

Author index